EVERYTHING'S RELATIVE

Also by Tony Rothman

Nonfiction

Doubt and Certainty (with George Sudarshan)
Instant Physics
A Physicist on Madison Avenue
Science à la Mode
Frontiers of Modern Physics

Fiction

Censored Tales
The World Is Round

EVERYTHING'S RELATIVE

AND OTHER FABLES
FROM SCIENCE AND TECHNOLOGY

Tony Rothman

WILEY

John Wiley & Sons, Inc.

Published by John Wiley & Sons, Inc., Hoboken, New Jersey
Published simultaneously in Canada

Permission to quote George Antheil's description of his meeting with Hedy Lamarr as written in his 1945 autobiography *Bad Boy of Music* on pages 179–180 is granted by the Estate of George Antheil.

For general information about our other products and services, please contact our Customer Care Department within the United States at (800) 762-2974, outside the United States at (317) 527-3993 or fax (317) 572-4002.

Wiley also publishes its books in a variety of electronic formats. Some content that appears in print may not be available in electronic books. For more information about Wiley products, visit our web site at www.wiley.com.

Library of Congress Cataloging-in-Publication Data:

Rothman, Tony.
    Everything's relative : and other fables from science and technology / Tony Rothman.
      p. cm.
    Includes bibliographical references and index.
    ISBN 0-471-20257-6
    1. Science—History—Miscellanea. 2. Technology—History—Miscellanea. I. Title.

Q125.R763   2003
509—dc21

                          2003013944

Printed in the United States of America

10 9 8 7 6 5 4 3 2 1

The common belief that we gain "historical perspective" with increasing distance seems to me utterly to misrepresent the actual situation. What we gain is merely confidence in generalization which we would never dare to make if we had access to the real wealth of contemporary evidence.

—O. Neugebauer, *The Exact Sciences in Antiquity*

CONTENTS

PREFACE

Napoléon once said, "History is a fable agreed upon." Perhaps it was "fraud." According to authorities, he made the remark after Waterloo when Wellington asked him how it felt to suffer history's most monumental defeat. Napoléon replied, "History is a fraud agreed upon," and was exiled.

This book is an amateur's stroll through the history of science. It is not a serious book. Professional historians will sniff that what I write is well known and that I have left out the really important, the agonizing minutiae. Professional scientists will shrug: "Who cares? None of this makes any difference to the course of progress anyway." Both are probably right. I thank the historians for their diligence, which I have plundered mercilessly. To the scientists I say, well, I am a professional, and if we cannot profit by the mistakes of our forebears, at least we can be entertained by them.

No, this book does not intend to edify professionals of either a scientific or historical persuasion. My position is closer to that of the cavalier who pursued the famous courtesan Madame Recamier. When she rebuffed his advances with the pretense "Monsieur, my heart belongs to another," he answered without a blush, "Madame, I never aimed so high as that."

My aim is at the rest of us, the masses weaned on high school and college texts, television and magazines. My weapon of choice is the anecdote, more precisely the antianecdote. Although serious work goes on in the history of science, by the time it filters down to the level of high school or college texts, NASA Web sites, *Time* magazine, or even the average popular science book, it has become a collection of just-so stories. You know what I mean: Edison invented the lightbulb, Morse invented the telegraph, Fleming discovered penicillin, George Washington never told a lie. When it comes to scientists, well . . . scientists occasionally work, but the really significant advances are made by solitary geniuses in blinding flashes of insight, usually when stepping onto a streetcar or dozing off before a fire.

So I am going to tell some stories, antistories. I hope one or two will surprise you. The reason I am going to tell stories is because the rage, current now for some years, has been for narrative. The trouble with narrative is that one gets carried away with the, um, narration. The fact is, most popular science books take two or three hundred pages to tell a story that might be reasonably condensed into about five pages, and that is because they are more concerned with biography, glamour, plot . . . narrative.

The facts. Yes, because I'm a writer I like to tell stories, but because I am a scientist I have a penchant for facts. So I'll tell a few stories, but I hope they will be reasonably well researched stories, antianecdotes. I'll keep them short, maybe a dozen pages per dose, so that if I'm lucky, there will be at least one page of content. Strangely, if you read from beginning to end, you will find that this hunchbacked history does seem to contain a narrative thread, if one told by the losers as often as the winners.

I have few illusions that this little book can make a dent in just-so history. Just-so history just holds too many advantages for all parties gathered around the campfire. It is, after all, perfect for textbook writers and journalists who, hamstrung by space limitations, need microwavable history to pass on to their readers. Readers and listeners are rewarded as well. Let's face it, we aren't a people interested in process. Results, that's what we need. In any case, these days we are overloaded. We lack sufficient ROM to deal simultaneously with nonfunctioning Web sites, customer nonservice, and ringing cell phones *and* to be forced to remember more than one hero, more than one discoverer, more than one version, a hint of subplot. How much more efficient to assign an invention to Bell, a discovery to Fleming, a dream to Kekulé, rather than to find out what really happened—or what really didn't happen.

Just-so history is perhaps most convenient for scientists themselves. Not only is it easy to remember, but it is simultaneously a history of logical completeness, for lack of a better term. From the blinding flashes (the only sort produced when dozing off before a fire on a streetcar) spring forth logically complete and completely understood concepts.

Just so.

You see, scientists make lousy historians. And for many years they were the only people who wrote the history of science. Whereas science is sometimes logical, history is not. Scientists, equipped with superior analytical faculties, are unsurpassed at reconstructing history into a seamless narrative to arrive at the present state of affairs. Anyway, why spoil a good story? After twenty or thirty years in the field, after hundreds of tellings, retellings, overtellings, and undertellings, after your logic and your perfect hindsight have laundered history until it is neat, pressed, and starched, are you going to ruin everything? Of course not. You *know* it must have happened like that.

If it strikes you as odd that the profession most vociferously devoted to Truth should remain unstimulated to apprehend the truth of its own past, you are not alone. I often wonder the same thing before dozing off on a bicycle in front of a streetcar. Actually, the evolution of just-so history from the point of view of the scientist is not too difficult to understand. Self-interest is at work here and a little self-aggrandizement. We partake of the heroes we create.

Forgivable, surely. Yet there is a downside—the fairy tales created by scientists have tended to career toward the exclusionary. In the old days, at least, they were too often designed to persuade the rest of the world that scientists are different and to dissuade potential scientists from joining the ranks; if you aren't among the most brilliant, don't apply. Once upon a time I published an article criticizing many of the biographies of Evariste Galois, a famous mathematician who died in a duel at the age of twenty. My thesis (which will not surprise you) was that the standard biographies of Galois, all written by scientists, were largely fabricated. This caused controversy. The *New York Times* didn't like it, complaining that I spoiled a legend. But one of my mentors, a famous physicist, wrote to me, "I agree with you completely about the cause of the falsification, the narcissistic self-glorification of infant prodigies grown old."

As I said, I have few illusions that this book will make the slightest puncture in the status quo. Nevertheless, I feel there are a couple of lessons to be learned from this meander down the back alleys and cul-de-sacs of science. The most important is

He Who Hath, Gets

What does this mean? It means, "The rich get richer." I illustrate with my one and only Richard Feynman antianecdote. It is nothing more than a footnote to a footnote to history, really, but it's all I have.

When Richard Feynman died, the physics trade journal *Physics Today* devoted the February 1989 issue to his memory. In it appeared a letter from Freeman Dyson to Sara Courant describing a conference that was held in 1980 in Austin.

> Dear Sara:
>
> I just spent a marvelous three days with Dick Feynman and wished you had been there to share him with us. Sixty years and a big cancer operation have not blunted him. He is still the same Feynman that we knew in the old days at Cornell.
>
> We were together at a small meeting of physicists organized by John Wheeler at the University of Texas. For some reason Wheeler decided to hold the meeting at a grotesque place called World of Tennis, a country club where Texas oil millionaires go to relax. So there we were. We all grumbled at the high prices and the extravagant ugliness of our rooms. But there was nowhere else to go. Or so we thought. But Dick thought otherwise. Dick just said, "To hell with it. I am not going to sleep in this place," picked up his suitcase and walked off alone into the woods. In the morning he reappeared, looking none the worse for his night under the stars. He said he did not sleep much, but it was worth it.

What really happened was this. At the time of the conference I was a graduate student at the Center for Relativity at the University of Texas, and John Wheeler was one of my professors. The meeting was by invitation only. I was not invited. Wheeler, however, gave me a lollipop: chauffeur Feynman, Dyson, and John Schwarz out to World of Tennis, which was located on the shores of Lake Travis (and may still be). The plan was to take them out one at a time, but due to flight delays I drove them all out together in a tiny car I had borrowed from a neighbor.

The evening was dreary, and it had begun to drizzle. During the hour-long drive one thought and one thought alone tortured me: *If I have an accident, physics has ended.* Actually, Feynman and Dyson were past their prime and physics would have continued its majestic course without them. But Schwarz, then an unknown postdoc, was several years away from his big superstring breakthrough. A crash might have aborted the theory of everything. Feynman was in a cantankerous mood, his usual. He pounced on anything anyone said.

We argued every minute. We arrived at World of Tennis and I took them into the lobby to make sure we had come to the right place. As we said good-bye Feynman turned to me and asked where I was staying.

I'm camping out by the lake," I replied.

"Let me come."

I waited and waited, but no sign of Feynman. It was still drizzling, and since Feynman had not appeared, I decided to head back to town. I have no doubt that he was looking for me when he abandoned his room. In any case, to camp outside was my idea. I claim priority.

A footnote to a footnote of history, as I said. On a larger scale such historical whitewashing is the rule. An MIT mathematician once remarked to me that he thought the work of his entire generation would be attributed to the great Russian mathematician Kolmogorov. We labor in vain. It is true. I weep.

Although he who hath gets, many should have gotten. Call it conservation of credit. I, honestly, have little interest in denying credit to anyone, but being of socialist inclinations, I prefer to see it redistributed.

This Principle of Redistribution is perhaps the easiest lesson gleaned from the history of science and follows from the most cursory dip into the past. For every famous musician, ten others are equally good. So it is with science. Unique discoveries are the exception rather than the rule. Discoveries almost always take place simultaneously, or nearly so. All scientists know this. To be sure, with the increasing pace of scientific research everything happens at once. It is quite clear that if you don't do something, someone else will. Someone else already has.

Virtually all discoveries have forgotten precursors: people who almost got it but did not see the implications, people who did get it but said it badly, people who lived in the wrong country, people who were so far ahead of their time as to be invisible. Following on the heels of the Principle of Redistribution is therefore the second supremely important law of the history of science, the Infinite Chain of Priority:

Somebody Else Always Did It First

The Infinite Chain of Priority has consequences for one's worldview. Just as in art, when learning how much Stravinsky stole from Debussy and Rimsky-Korsakov forces a reevaluation, so too in science, when uncovering the hordes of forgotten precursors forces one to the conclusion that there are no revolutions, only evolutions. With perhaps the exception of Einstein and one or two others.

We remember only he who carries the torch past the finish line. But, unlike a racetrack, the course of science is not straight, or even circular. As every researcher knows:

You Only Get the Right Answer After You've Made Every Possible Mistake

Arthur Koestler might call this sleepwalking. Let us term it the random walk of research, of life. So it is in the small, so it is in the large: Not only must individual researchers make all possible mistakes, but history itself must explore all possible dead ends before it finds the route across the finish.

To be sure, the torchbearer's torch is an optical illusion. In the Contemporary Panopticon of Present and Past Concepts (step this way, please), one principle was until recently completely absent from scholarly history: advertising. Yet a well-known maxim in science is:

Either You Do the Calculation or You Get the Credit

This is also known as the Zel'dovich Principle, after the famous Russian astrophysicist Yakov Zel'dovich, who frequently advised, "Without publicity there is no prosperity." Carl Sagan was well aware of the Zel'dovich Principle. Sometimes you encounter it as "Either you do the experiment or you get the credit," or, more vaguely, "Either you do the work or you get the credit." (Variations are easy to construct.) In any case, the implications of the Zel'dovich Principle are clear: You can be either a researcher or a publicist, but time and life rarely permit both. It is an unfortunate fact, but careers in science are not always built on achievement; the reverse holds as well.

Yakov Zel'dovich, by the way, discovered everything before anyone else—at least that is the impression you get reading his papers. Yakov Zel'dovich? Who? Most Americans have never heard of him. And that leads to another principle, an unfortunate one:

The Colors of the Flag Obscure the Sun

I mean nationalism. It may seem preposterous on the face of it that all the great inventions of the nineteenth and twentieth centuries should have been made in the United States. Actually they weren't, but whether you are talking about the airplane, television, lightbulb, or almost anything else, our countrymen get the credit. If you lived in Russia or Germany or even Uzbekistan (well, maybe not Uzbekistan), the same would apply. But the fact is that intelligence, imagination, and industry recognize no borders.

These are a few of the lessons that will be gleaned from this book. We now abandon the sound-bitten histories of textbooks and *Time* magazine and prepare to stumble into the lapses and gaps of times gone by and forgotten. When we emerge we find that science is done by human beings, often brilliant, sometimes bumbling, occasionally nefarious. It is the simplest lesson of all, but, like the history itself, the one most frequently forgotten.

LAPSES, SOURCES, AND
ACKNOWLEDGMENTS

The contents of this book are entirely arbitrary, having been determined by accident. I have endeavored to avoid sweeping generalities and the setting up of straw men. You won't find here such declarations as "Everybody believes George Washington never told a lie, but that isn't true." Well, duh. Neither, I trust, have I spent too much time on trivialities: "George Washington's false teeth actually weren't ivory but wood." (I am shocked.)

At the same time I have avoided topics about which there is no contention. You will not find a chapter about how the Wright brothers really *did* build the first powered airplane that successfully carried a person. I have sidestepped ancient science, "ancient" being more than about two thousand years old, in that there is no point in taking issue with stories known to be legendary. Somewhat arbitrarily, I have opened the book with the dawn of modern science—Galileo—and even those tales are chestnuts. With no less arbitrariness, I have closed the technology section with a chapter on secret electronic communication. I am well aware of the dispute over the more recent origin of the computer, but space limitations required cutting a chapter on this topic. Readers who are unfamiliar with the story can begin researching it by doing a Web search on "Atanasoff." I have also posted the deleted chapter on my Web site, which is most easily accessed by doing a Google search on my name.

This raises the issue of one important source: the Internet. The Internet has proved to be both a boon and a curse. It is a galaxy wide and an inch deep. You can find something about almost anything on the Net— including almost all the topics in this book—so why bother writing at all? (My students seem to have adopted this philosophy.) Unfortunately, the overwhelming majority of Web sites do a poor to nonexistent job of documenting their sources and giving basic citations. As a typical example, a number of sites list William Roberts as the discoverer of penicillin in 1874 (see chapter 19), but *none* reference the source of this information or the context. On a hunch I checked the *Philosophical Transactions of the Royal Society* for that year. Sure enough, a Dr. William Roberts did report a few observations concerning penicillin, but he was largely following up some rather famous experiments done even earlier by Sir John Burdon-Sanderson, who is conspicuously not mentioned on the same Web sites. And so it goes; the Internet has become history's largest rumor mill, where false claims get multiplied thousands of times over and establish their veracity by weight of numbers. Any site should be surrounded by warnings: Danger, Weak Link.

On the other hand, every volume of the *Philosophical Transactions of the Royal Society* ever published (from 1665) is how available online, and this is a boon. Thus, I have often used the Internet as both a foil and an entry to research, but if there is one lesson to be learned from any historical sleuthing, it is *read the original*. Unfortunately, originals are often impossible to obtain, and I have frequently relied on secondary sources. The accuracy of my claims will depend on my judgment as to their reliability. It's too much to hope that I will have gotten everything right, but at least I have listed my sources. Irate readers should consult them and come to their own conclusions. Because this is a popular book, I have not wanted to clutter it up with too many footnotes, but my sources, keyed to the relevant page numbers, can be found with some in-depth commentaries at the back of the book.

Finally, the ideal author of a book such as this would read half a dozen different languages. The real author, unfortunately, is restricted to two and a half. Inés Arribas provided great help in checking my translations from French, and Imke Meyers was gracious enough to plow through a sheaf of old German papers with me. Thanks also are due to Tony Heaton at Illinois Wesleyan University's Interlibrary Services for putting up with my unceasing stream of strange document requests. For much the same reason I am heavily indebted to the librarians at Bryn Mawr College, especially to Andrew Patterson and Anneliese Taylor. Marc Ellis, the editor of the *Old Timer's Bulletin,* has provided me with entrees into the literature of early radio and television. At the Moscow Polytechnical Museum I thank Roman Artemenko and Sergei Shevchenko for personally guided tours. Much gratitude is due to Hans Bebie and Viktor Gorgé at the University of Bern for a time-consuming, if futile, search for evidence that Einstein attended seminars there. Others who have provided documents or pointed me in the right direction are Pierre Laszlo, Max Lazarus, Albrecht Wagner, Mauro Piccinini, Jost Lemmerich, Peter Rowlands, Herman Kruis, George Dyson, and Gabe Spalding. My thanks to them all.

I

The Domain of Physics and Astronomy

The Contemporary Panopticon of Present and Past Concepts can be viewed in many ways. Unlike conventional museums, there is no fixed floor plan or identifiable structure. The route depends entirely on the whim of the visitor. One may take a chronological stroll through the exhibition area, attempting to link the development of scientific concepts into one long, unbroken chain from past to present. Alternately, one may choose the biographical route, following personalities as they parade across the Panopticon's infinite facade. One may jump from physics to chemistry to medicine to technology at the blink of an eye. All routes are encouraged. Only gradually does one begin to perceive the extent of the Panopticon's holdings.

For today's introductory tour, the Panopticon has been conveniently divided into domains containing related subjects, and these have been laid out in chronological order. (However, as remodeling is ceaseless, one should take care to avoid the holes in the floor and ceiling, and ignore the large number of exhibits shut down periodically for renovation.) The Domain of Physics and Astronomy is the oldest at the Panopticon, and as such contains the largest collection of dubious anecdotes, misattributions, and outright legends.

We cannot pause at all of them. Take a moment to glance into the extensive exhibit on legends. Most of the stories of the ancient Greeks find a place there. We do not set foot on Ionia; instead we highlight two displays from the dawn of modern science. These are the old chestnuts that have Galileo publicly dropping weights from the Leaning Tower of Pisa (thus proving that all objects fall at the same rate, regardless of their mass) and discovering the principle of the pendulum by timing the swing of a lamp in the cathedral at Pisa with his pulse. Nowhere in Galileo's writings will you find a word about either experiment. Surprised? The Panopticon is filled with imaginary and doubtful demonstrations. You might want to stop by Young's experiment in chapter 2.

Both Galileo stories originated with his pupil Vincenzio Viviani (1622–1703), whom we shall meet in the flesh shortly, and who published the *Life of Galileo* in 1717. (Rather, one should say it was published for him, as Viviani, being a perfectionist, kept polishing it for fifty years until he died, whereupon the matter was taken out of his hands.) This first biography of Galileo was consciously patterned after the lives of the saints. To Viviani, his master was the ideal Renaissance genius, which may explain why we refer to Galileo alone among scientists by his first name, along with Leonardo, Michelangelo, and Raphael. Indeed, Galileo was born three days before Michelangelo died, in 1564, and there is substantial evidence among Viviani's notes that he was attempting to find a transcendental link between the two. Myth and history are so intertwined in the pages of the *Life* that no one will ever be able to sort them out.

In regard to the Leaning Tower, the entire description of the experiment takes two lines. Galileo was engaged in controversies with Aristotelian philosophers; he attempts to test Aristotle's idea that heavier objects fall faster than light objects but finds the contrary,

> demonstrating this with repeated experiments from the height of the Campanile of Pisa in the presence of the other teachers and philosphers and the whole assembly of students.

Not a word more or less. In particular, no data. Nevertheless, even in our own time reputable Galileo scholars have taken Viviani at his word.* No one will ever be able to refute them, but to repeat, Galileo himself never mentions the experiment. To the extent that in unpublished notes Galileo discusses dropping lead and wood weights from towers—"This is something I have often tested"—he claims that the lead weights "move far out in front." In other words, rather than refute, he *verifies* the Aristotelian doctrine that heavier objects fall faster than light objects.[†] Many students, by the way, uphold this doctrine today—until they are purged of Aristotelian notions in their first physics course. Ontogeny recapitulates phylogeny. Letters as late as 1641 from Galileo's colleague Vincenzio Renieri indirectly indicate that at one time Galileo indeed supported Aristotle on motion. Whatever. At least someone performed the experiment: Simon Stevin of Bruges (1548–1620) did drop weights from a tower before 1605 and, unlike Galileo, arrived at the correct answer.

---

* For instance, Stillman Drake and Giorgio de Santillana.
† What exactly Aristotle believed on this score is itself the subject of debate (see Lane Cooper, *Aristotle, Gailileo, and the Tower of Pisa* [Ithaca: Cornell University Press, 1935]), but it is fairly certain that he did not believe what we believe today.

*　　*　　*

Unverifiable stories abound in the Panopticon; for this reason the Leaning Tower is sometimes found in the Historical Controversies sector of the museum. Take a moment to glance at the more modern story of superconductivity. Superconductivity, the ability of certain materials to conduct electrical currents without any resistance whatsoever, was one of the great discoveries of the twentieth century. Check any encyclopedia (even books about the Nobel prize) about who discovered it, and you will find it attributed to Heike Kamerlingh Onnes in 1911, a discovery for which he won the coveted award. In fact, Kamerlingh Onnes won the prize for the liquefaction of helium. Judging from reports by those close to the work, there can be no question that it was an assistant, Gilles Holst, who first observed that near absolute zero the element mercury lost all electrical resistance. It is also true that Holst's name never appeared on the paper. As to the rest—rumors about injured pride, about professional misconduct, and that Kamerlingh Onnes initially disbelieved Holst's results and had to be persuaded by repeated trials (they all did)—all this seems to be unverifiable.

As we move on, note to the right the entrance to one of the largest wings at the Panopticon: the Hall of Misattribution. Almost anything you can think of is named for the wrong person. The famous Aharonov-Bohm effect in physics, to take just one example (we do not pause to explain it, but it is famous), was proposed ten years earlier in a paper by Ehrenberg and Siday that appeared in a prominent journal. The effect, however, was explained only at the end of their rather long article, which may mean it is wise *to get to the point*.

With this in mind, we allow you to browse the Hall of Misattribution at your leisure. Let us now return to the time of Galileo and inspect one of the most famous stories of all: the invention of the barometer.

1 / The Mafia Invents the Barometer

Textbooks mention half of him: "Other units [for pressure] in common use are the atmosphere, the millimeter of mercury, or torr, and the millibar." Gads. It is a damnation of the highest caliber: He has been magnified from a person to a unit and lost his name. Truncated and lowercased, proof positive that he has faded into the cultural background like his invention, which hangs uselessly on the walls of seafood restaurants. Sometimes writers do let drop his entire name. The reference is invariably laconic: "Another instrument used to measure pressure is the common barometer, invented by Evangelista Torricelli (1608–1647)." Air pressure, barometer. Ah. Once in a great while, when an author turns reckless, Torricelli flickers momentarily in human form. Berte Bolle, from his history of the barometer, bravely:

> Torricelli set up his tube of more than 33 feet (10 meters) long in his house with the top protruding through the roof. He floated a small wooden dummy on the water at the top of the tube; in bad weather the height of the water column fell so much that the dummy could not be seen from the road whereas in fine weather it floated high and clear for all to see. It was soon rumoured that master Torricelli was in league with the devil and the water barometer was quickly removed!

We are convinced. But wait. In Sheldon Glashow's account, Torricelli carries on his heretical work, darting around the quayside to the delight of onlookers. Rumors, evidently—and the Inquisition—failed to deter him: "Torricelli filled long tubes, sealed at one end, with liquids such as honey, wine and seawater, and lashed them upright to ships' masts. He found that the height of the column depended only upon the total weight of the liquid within."

Isaac Asimov, eschewing drama for knowledge, provides a complete tale for his readers' edification. The immortal Galileo, Torricelli's boss, suggested that his assistant investigate why water pumps failed to raise water more than ten meters above its natural level. Those were the days. Science was called philosophy, Aristotle held sway, and Nature abhorred a vacuum. Galileo's position was purely Aristotelian: Pumps create a partial vacuum above the water, and the water rushes in to fill it. The vacuum sucks. Evidently, however, the vacuum's ability to suck had limits—about ten meters. Asimov relays Torricelli's thoughts:

> It occurred to Torricelli that the water was lifted, not because it was pulled up by the vacuum, but because it was pushed up by the normal pressure of air. After all, the vacuum in the pump produced a low air pressure, and the normal air outside the pump pushed harder.
>
> In 1643, to check this theory, Torricelli made use of mercury. Since mercury's density is 13.5 times that of water, air should be able to lift it only 1/13.5 times as high as water, or 30 inches. Torricelli filled a 6-foot length of glass tubing with mercury, stoppered the open end, upended it in a dish of mercury, unstoppered it, and found the mercury pouring out of the tube, but not altogether: 30 inches of mercury remained, as expected.

Admirable detail. We feel as if we are face-to-face with Torricelli. "Hand me the mercury," he says. Totally irreconcilable, then, is this remark retrieved from cyberspace: "In 1643, Torricelli proposed his experiment, which was carried out by his colleague Viviani."

Detail. Ah.

The truth is, no one is entirely sure what happened. We do know they were Italians and they were friends. Today they would form a research group. When the research group monopolizes a territory we call it a mafia. Then, as now, the senior scientist receives the credit. To understand what no one is certain about, we return to the dawn of the seventeenth century. The Counter-Reformation in Europe is under way, the Inquisition is heating up, Galileo condescendingly ignores Kepler's discovery that planetary orbits are ellipses rather than circles, Newton has yet to be born. On the ground, the outstanding philosophical question of the age boils: Is a vacuum possible?

No. The answer is obvious; let's be off to today's witch trial. That, at least, is the current universal opinion, nineteen hundred years old. Any objection will be met by a citation from the supreme authority, Aristotle. Aristotle, in his celebrated phrase, declared, "Nature abhors a vacuum." (Whatever Aristotle declared, he declared in ancient Greek, but this is how it is usually translated, and he believed it.) Aristotle adduced a number of arguments against the vacuum, both physical and logical. You must first understand that in Aristotle's world—and in the world of the sixteenth century—there are no atoms. Water is a continuous substance. Dividing water into finer and finer pieces leads only to finer and finer pieces, ad infinitum. There is no reason to suppose that the division will lead to a state composed of ultimate particles between which is nothing. No, the universe is full, a plenum. What is more, in the pre-Galilean world, there is no concept of inertia, the idea that without interference an object travels at a constant velocity. Rather, the velocity of an object depends on the *resistance* of the medium through which it travels. A void— a vacuum—provides no resistance. Therefore the velocity of an object traveling through a void should be infinite. This is clearly nonsense.

Those are physical arguments that Aristotle brought against the vacuum. His main logical argument was that the position of an object—its place—is always understood to be within the inner limits of a surrounding body. Nonphilosophers call this a container. But the void has no properties. An object within it cannot be said to be in any sort of place. Neither could an object be said to move within a void (because it has no properties to distinguish places). Therefore an object cannot have a place unless it is within some substance. A vacuum is logically impossible.

If a vacuum is logically impossible, that would mean that God could not produce one if he wished. This troubled thirteenth-century theologians. For that reason, by the seventeenth century people were willing to discuss the issue. Yet the prevailing opinion was that a vacuum was at least a physical impossibility, if not a logical one.

In the Contemporary Panopticon of Present and Past Concepts, the exhibits on vacuum and pressure are housed side by side. This way, please. From our perspective, it is difficult to see how a sensible concept of vacuum could emerge without a sensible concept of pressure. An anonymous thirteenth-century pupil of the philosopher Jean de Némore understood that pressure in a liquid increased with depth, but the publication of Némore's book in which the discussion appears was delayed for three centuries. Isaac Beeckman (1588–1637) seems to have accepted the idea of a vaccum and in 1614 wrote in his journal that air has weight and exerts pressure on bodies below, which increases with the depth of the air. Despite such isolated beacons of insight, a clear understanding of pressure was not to be had. Air is weightless.

Two years before Beeckman grasped the essentials, Galileo, in a fit of pique, expressed this universal wisdom: "Even if we then add a very large quantity of water above [the solid], we shall not on that account increase the pressure or weight of the parts surrounding the said solid." A year after Beeckman, in 1615, Galileo continued his denials: "Note that all the air in itself and above the water weighs nothing. . . . Nor let anyone be surprised that all the air weighs nothing at all, because it is like water."

Against this background Giovanni Batista Baliani (1582–1666), from Genoa, wrote to Galileo in 1630 to report the results of an experiment.

He had attempted to siphon water from a reservoir over a hill about twenty-one meters tall, and the siphon failed to perform. The siphon, in a procedure known to gasoline thieves today, was initially filled with water and laid over the hill, but when the tube was unstopped, the water level on the reservoir side dropped back to about ten meters. Mystery? Not to Galileo. He condescended to reply to Baliani that the answer was obvious: The force of the vacuum raised the water, but the strength of the vacuum was limited to ten meters. Baliani was closer to the mark: He believed that a vacuum was possible and that water and air had weight.

He also had friends. Of the right sort. They included Raffaello Magiotti, Evangelista Torricelli, Emmanuel Maignan, Athanasius Kircher, Niccolò Zucchi, and, evidently, Gasparo Berti. This was the Roman mafia. Somewhere between 1639 and 1641—the dates have been eliminated—Berti performed an experiment at his house in Rome. The mafiosi Kircher, Magiotti, and Zucchi were there; Maignan was absent; and Torricelli's whereabouts are unknown. Four accounts exist of the experiment, three by the eyewitnesses and one by Maignan, who was informed of the proceedings by Berti a week later. The accounts differ on the details; over the interpretation of the results they came to blows.

According to Maignan, "one of the keenest minds of the seventeenth century," the experiment was set up roughly as follows. Berti clamped a long leaden tube, at least "forty palms" in height, to the outside of his house. The bottom of the tube, which ended in a barrel of water, was fitted with a valve. Over the top end of the tube was sealed a glass flask, which was also fitted with a stopcock. The experimenters closed the bottom stopcock, then from a tower window filled the entire tube, including the glass flask, through the upper valve. The upper stopcock was closed, the bottom one opened.

Tension. Suspense.

The water level falls—but not completely. The experimenters lower a sounding line into the tube to determine the height of the water. The data are in: eighteen cubits. This is the height to which Galileo claims an air pump can raise water. The water level stands for a day. The experiment is repeated with variations. The data are solid. But what is the space above the water? When the philosophers first opened the upper stopcock to lower the sounding line, they heard a loud noise as air rushed in. Air rushing in—that is Maignan's view. The fall of the water level in the tube therefore must have left a vacuum behind. Fellow mafiosi are unconvinced.

The plenists argue that air seeped in through the pores of the lead or the glass in order to fill up the space left by the falling water. Kircher, apparently, suggests putting a small bell into the glass bulb and attracting the clapper to one side with a magnet. If within the flask exists a vacuum, no sound will be heard. Maignan objects that the glass itself will conduct the sound, and no documents in the Panopticon make clear whether the experiment is ever carried out.

Today the breakthrough would have won a Nobel prize. Then, news was kept in the family. They were a congenial bunch, judging from the letters among them, reveling in the vistas of the Golden Age that opened before them. They may have also held doubts about the Inquisition. Vacuums, you know. In 1648, some years after Berti's experiment, Raffaello Magiotti, who was there, wrote a letter to Father Mersenne in Paris, mentioning that he had told Torricelli about Berti's tube and that "they" had since made many demonstrations with quicksilver. *They.*

The mercury connection. Torricelli, born on October 15, 1608, had attended the University of Rome and had become a recognized mathematician. They say he was charming. By the end of 1641 he had become Galileo's assistant, but Galileo died only three months later, to be followed by Torricelli himself in 1647. In the meantime Grand Duke Ferdinand II made Torricelli philosopher and mathematician in Florence, a joint appointment rarely encountered today. He remained in Florence, publishing until he perished, we hope in better circumstances than Galileo.

The idea for using mercury in a device similar to Berti's may have come from that archfoe of air pressure, Galileo (perhaps he had repented). In a copy of the original edition of Galileo's *Discorsi* of 1638, there appears a marginal note made in the hand of his assistant of the time, Vincenzio Viviani, "with the approval of Galileo himself." The note reads, "It is my belief that the same result will follow in other liquids, such as quicksilver, wine, oil, etc., in which the rupture will take place at a lesser or greater height than 18 braccia, according to the greater or lesser specific gravity [density] of these liquids in relation to that of water." Viviani is a great friend of Torricelli. Ah.

Events become obscure. The first full account of Torricelli's famous experiment, described by Asimov and Bolle in hyperrealistic detail, comes nineteen years after the fact. In 1663, one Calo Dati, a pupil of Torricelli, pseudonymously published letters from Torricelli to his best friend, Michelangelo Ricci, who may also have been present at Berti's experiment. These letters report the first experiments with mercury, that is, the barometer.

On June 11, 1644, Ricci wrote to Torricelli, "I live in a great desire to know the success of those experiments that you indicated to me." Torricelli penned his celebrated reply the same day:

I have already hinted to you that some sort of philosophical experiment was being done concerning the vacuum, not simply to produce a vacuum but to make an instrument which might show the changes of the air, now heavier and coarser, now lighter and more subtle. Many have said that [the vacuum] cannot happen; others say that it happens, but with the repugnance of nature.

Torricelli goes on to espouse his own view that the vacuum is not the issue and that one can be produced. Then the immortal phrase *"Noi viviamo sommersi nel fondo d'un pelago d'aria elementare"*:

We live submerged at the bottom of an ocean of elementary air, which is known by incontestible experiments to have weight, and so much weight that the heaviest part near the surface of the earth weighs about one four-hundredth as much as water.

He goes on to say, "We have made many glass vessels . . . with necks two ells long." *We.* The tubes, closed at one end, were filled with mercury, so that no air remained at the closed end, then inverted in a basin of mercury; as Asimov describes, the mercury falls, but not completely. Torricelli clearly understands that it is not the vacuum exerting an insufficient force on the quicksilver:

I assert . . . that the force comes from outside. On the surface of the liquid in the basin presses a height of fifty miles of air; yet what a marvel it is, if the quicksilver enters the glass [tube] . . . it rises to the point of which it is in balance with the weight of the external air that is pushing it! Water, then . . . will rise to about eighteen ells, that is to say, much higher than the quicksilver, as quicksilver is heavier than water, in order to come into equilibrium with the same cause, which pushes the one and the other.

Thus, a thoroughly modern understanding of air pressure and the invention of the barometer, which measures that pressure. A more modern understanding than modern English usage would indicate: We do not suck soda through a straw; air pressure pushes it into our mouths.

But: "We have made many glass vessels." *We.* According to Dati, who, as we know, first reported the experiment nineteen years after the fact,

Torricelli did not perform it. He forecast the result to Viviani, who procured the mercury, had the apparatus built, and verified his friend's prediction. Thus an early example of a familiar division of labor, theorist and experimentalist.

What of Torricelli's dockside activities, lashing glass tubes filled with water and wine to the masts of tall ships? That appears to be a confusion with Blaise Pascal, who performed such demonstrations in 1647 to the delight of the French public—at the Rouen glass factory. Thus were forever bound together the three sensual delights, wine, water, and barometers.

Pascal, they say, wrote to his brother-in-law Florin Perier and suggested that he take a barometer up Puy-de-Dôme to test whether the weight of the air varied with altitude. Descartes also claims priority for the idea, and textual analyses indicate that the letter from Pascal to his brother-in-law may indeed have been a falsification. Whatever. On September 19, 1648, Perier did climb the mountain. The height of the mercury in the barometer fell. No longer was there any doubt: Air pressure varied with height. The vacuum was abandoned, in horror. It is true: We live submerged at the bottom of an ocean of elementary air, which is known by incontestable experiments to have weight.

2 / The Riddle of the Sphinx: Thomas Young's Experiment

As one scans the Contemporary Panopticon of Present and Past Concepts for crucial turning points in the history of science, no more important experiment stands out—in classical physics, in quantum mechanics, or perhaps in science altogether—than the celebrated "double-slit" experiment of Thomas Young. It was, any textbook will tell you, carried out by Young in the year 1800. It was announced by him before the full assembly of the Royal Society of London. And it was the experiment that conclusively proved, after a century of debate, that light was not composed of particles, as Newton had believed, but was a wave.

To this day, every freshman physics major repeats Young's experiment in the basement lab. One shines a laser beam through two narrow slits separated by a fraction of a millimeter, slits that have been etched into an opaque slide. On a distant wall, instead of a single beam, one sees a series of light and dark bands—interference fringes, they are called. The student measures the distance between the slits, the distance from the slits to the wall, and the distance between two of the bright bands, and—*voilà!* Multiplication and division on a pocket calculator yield the wavelength of laser light to an accuracy of a few percent.

It is a great experiment, full of explanation and mystery. To pursue Young's experiment to its full depths takes you to the very heart of quantum mechanics and to the most fundamental questions about the nature of reality. The only question is, did Young do it?

Thomas Young was one of the great prodigies of his age. Born into a Quaker family on June 16, 1773, in the English village of Milverton, Thomas (named after his father) was the first of ten children. The others followed with a rapidity that forced Thomas to spend much of his first seven years in the home of his maternal grandparents, Robert and Mary Davis. This proved to be to his advantage. The Quaker tradition, with its emphasis on industry and the individual's search for truth, had and continues to contribute more than its share of outstanding scientists, Arthur

12

Eddington and Roger Penrose being two prominent names from the twentieth century. Young's grandparents, admonishing their charge that "a little learning is a dangerous thing / Drink deep, or taste not the Pierian spring," encouraged Thomas' curiosity in every direction.

The results, through nature or nurture, were astonishing. By Young's own account, written in Latin, he was reading "with considerable fluency" by the age of two and had gone through the Bible twice by the age of four. He was reciting poetry from memory by the age of five and began Latin at six. His next years were spent at several boarding schools. By the time Thomas left the Compton School in Dorsetshire at the age of thirteen, he had knowledge of Greek, Latin, French, Italian, Hebrew, and natural philosophy. These supplemented practical skills in the use of the lathe, lens grinding, and telescope making, crafts he had learned on the side from an assistant at Compton. Although Compton marked the end of his formal education for six years, Young continued to study on his own. In particular he read Newton's *Principia Mathematica* and *Opticks* and extended his repertoire of languages, which soon touched on Chaldean, Syrian, and Samaritan, and later Persian, Arabic, Turkish, and Amharic. On his deathbed he was compiling an Egyptian dictionary. Thomas Young, in fact, became one of the great linguists of the nineteenth century.

The secret of Young's success? His near-contemporary Poor Richard might have advised as Young once did his brother: "If you are careful of your vacant minutes, you may advance yourselves more than many do who have every convenience afforded them." On his deathbed, Young told a friend that his greatest satisfaction would be to have never spent an idle day in his life.

Despite his success with languages, Thomas felt it was more a matter of diligence than aptitude, and at any rate, a few languages were considered a normal part of a sound classical education. He opted to become a doctor and set off for London in 1792, at the age of nineteen. The following year he read his first paper, "Observations on Vision," before the Royal Society. Based on the dissection of an ox's eyeball, Young concluded that the lens was responsible for focusing images; the eyeball did not change its length, as others maintained. His paper had two immediate consequences, a priority dispute with another philosopher and Young's election as a fellow of the Royal Society, which took place in 1794, when he was twenty-one. There followed further medical studies in Edinburgh and Göttingen, where he wrote a dissertation on the human production of sound. In 1797 he returned to England and to Cambridge University in order to get an M.D. He remained there for two years, apparently more to satisfy the residency requirements than to read medicine, and returned, M.D., F.R.S., to London in 1799, where he prepared his first paper on the nature of sound and light for the Royal Society.

Such papers were no novelty at the Royal Society. Speculation on the nature of light and vision had a long and sometimes honorable history that trailed off into the mists of antiquity, and all parties could cite the Greeks in their favor. By the second half of the seventeenth century, two broad classes of theories had asserted themselves: particle theories and wave theories. The former, associated with the name of Isaac Newton (although he did not originate the idea), held that light was composed of minute particles—corpuscles, Sir Isaac called them. The latter, associated with the names René Descartes, Christiaan Huygens, Robert Hooke, and Leonard Euler, held that light was transmitted much as sound—by waves.

The incompatible views provoked a fierce and often uncivil controversy, reminiscent of the debate over the vacuum that had exercised philosophers a few generations earlier until the mafia put an end to it. Light was a harder nut, or perhaps corpuscle, to crack. Observations were plentiful, but due to the lack of any fundamental theory of matter, there was little to favor one explanation over another.

We do not pause to go into the details of the various theories, which are myriad and all contained in the Panopticon. Both wave and projectile theories could explain why light usually travels in straight lines, and both could explain the law of reflection—the angle of reflection equals the angle of incidence, as every pool player knows. Refraction, the bending of light as it passes from, say, air to water, is the phenomenon that causes pennies in a pool to appear in the wrong place and could be explained fairly naturally by the wave theory, but with somewhat more work, the corpuscle theory could also do the job.

The decisive test between the rival theories appeared to lie in the arena of phenomena today termed diffraction and interference. A Jesuit priest, Father Grimaldi, in a paper published posthumously in 1665, noticed that light passing through a narrow slit actually diverged more than it should if the rays were merely traveling along straight lines. True; if, say, a razor's edge is placed in a light beam, the rays are bent into the shadow region. What's more, the shadow is not sharp; a series of faint colored bands—Grimaldi's fringes—appear at the edge of the shadow region. Diffraction.

Also in 1665 Robert Hooke published a work, *Micrographia,* that described the colors produced by thin films or plates—for example, the colored rings you see in soap bubbles or on the ground when oil is spilled. With this explanation of the colors, Hooke introduced in a crude form the concept of interference (more in a moment). About seven years later, Newton himself performed experiments to investigate the films and reported the results in the *Opticks,* first published in 1704. He had placed a convex lens on a flat plate of glass so that there was a slight air space between them, a gap that grew wider from the point of contact outward. Shining white light down on the lens, Newton observed a series of concentric rings of different colors. The colors of Newton's rings—as they are called to this day—roughly followed the colors of the spectrum when viewed from the bottom and the complementary colors when viewed from the top.

Newton provided an elaborate explanation of the rings in terms of corpuscles, which we will avoid. As Young himself later remarked, "The mechanism . . . is so complicated and attended to by so many difficulties that the few who have examined them have been in general entirely dissatisfied." From today's perspective Newton's explanation does seem an exercise in futility, but as the Panopticon's exhibit on the Death of Theories shows, it is impossible to interpret an observation in an unbiased way, devoid of some conceptual framework. Our own outlook is so highly prejudiced, colored against Newton's, that it is very hard to make sense of his ideas.

Nevertheless, Newton was not as die-hard a corpusculist as universally believed. Early on, especially, his views were much more complicated and much more tentative. His frosty reply to archrival Robert Hooke's criticism of his 1672 paper makes this clear: "'Tis true, that from my theory I argue the corporeity of light; but I do it without any absolute positiveness, as the word *perhaps* intimates; and make it at most but a very plausible *consequence* of the doctrine, and not a fundamental supposition." Roughly speaking, Newton's theory required light to have both particle and wave aspects, but eventually he threw his weight behind projectiles, and such was his name and authority that there the matter rested for nearly a century.

Until Thomas Young cleared up everything.

Thomas Young's great contribution to the science of optics was to vigorously champion the wave theory and to put in a form comprehensible to us, his descendants, the principle of *interference.* Today interference is regarded as virtually the defining property of a wave. Two identical waves will pass through each other unscathed, but their heights, or amplitudes, combine. In places where crest meets crest, the waves are in phase and the waves interfere constructively, with a resultant amplitude twice that of the original. Where crest meets trough, the waves are out of phase, interfere destructively, and cancel out. The crucial point is that interference is a property of *waves;* corpuscles do not interfere. At the dawn of the nineteenth century, interference in water and sound waves was accepted. Young, reasoning by analogy, intended to prove that interference of light explained everything Newton could not.

As an example, take Newton's rings. Light passing downward through the lens is partially reflected upward at the lens's bottom surface. The remainder of the beam proceeds through the air gap and is reflected upward by the glass plate. If the color of light is associated with the wavelength of a wave (the distance between two adjacent crests) and the extra distance traveled by the second beam is an exact multiple of one wavelength, the two beams will be in phase. One will observe a bright ring of that color. On the other hand, if the extra distance is an exact multiple of one-half a wavelength, the two beams will be out of phase, resulting in a dark ring corresponding to that color. Clearly, as the gap between the lens and plate widens, different wavelengths will be in and out of phase, resulting in rings of different colors.

A parsimonious explanation, to be sure.* Yet how could one *prove* the wave nature of light?

The double-slit experiment.

Virtually every freshman physics text contains a schematic diagram of the setup, which is much as already described in the introduction. Young put a hole in an opaque screen to collimate (make parallel) the light rays. The collimated beam was then passed through two pinholes or slits and projected onto a distant screen. To reach a given spot on the screen, light from, say, the top pinhole must travel slightly farther (or less far) than light from the bottom pinhole. As before, if this extra distance is a multiple of one wavelength, the two beams interfere constructively, and a particularly bright spot results at that point on the screen. If the extra distance is a multiple of one-half the wavelength, the interference is destructive and the viewer observes a dark spot on the screen. Simple geometry makes it possible to calculate the wavelength of light from the distance between fringes, the distance between the slits, and the distance between the slits and the wall.

---

* There is actually a subtle problem with Young's explanation—that the center spot is dark. We let you puzzle it out.

This is all very impressive—it actually works—and even more impressive is that Halliday and Resnick's *Fundamentals of Physics,* the most famous undergraduate physics text of them all, reproduces in at least three editions "Young's original diagram showing interference of overlapping waves. (From Thomas Young, *Phil. Transactions,* 1803)."

This is all very mysterious. Often the year is 1800; Asimov gives 1801; 1802 is not ruled out. Inevitably, the citation is to one of Young's lectures before the Royal Society. Vain hope! You can scour the *Philosophical Transactions* for these years as closely as you like; you can read them upside down and backward or translate them into code. There is no mention of the double-slit experiment. Young's 1800 paper is devoted mostly to the wave nature of sound. Toward the end he argues by analogy that the nature of light is also wavelike, and he points out several deficiencies in Newton's theory. The 1801 Bakerian Lecture, "On the Theory of Light and Colours," begins with these stirring words:

> Although the invention of plausible hypotheses, independent of any connection with experimental observations, can be of very little use in the promotion of natural knowledge; yet the discovery of simple and uniform principles, by which a great number of apparently heterogeneous phenomena are reduced to coherent and universal laws, must ever be allowed to be of considerable importance toward the improvement of the human intellect.

Young then sallies forth with a vigorous argument for an "undulatory" theory. He is not above quoting Newton at length to demonstrate that the supreme authority was not wholly opposed to the idea of waves. In the Bakerian Lecture one can appreciate Young's gracious writing, and you will find there a detailed explanation of Newton's rings in terms of interference, much as I have explained it here. You will also find Young's precise values for the wavelengths of light inferred from Newton's own measurements: extreme red, "37640 undulations in an inch"; extreme violet, "59750 undulations in an inch." But you will not find the double-slit experiment. Neither will you find it in his lecture of 1802, where he enunciates very clearly the "law" of interference, nor in the Bakerian lecture of 1803, in which he recounts his repetition of Grimaldi's experiments and reports extremely precise results.*

---

* The law of interference: "Wherever two portions of the same light arrive at the eye by different routes, either exactly or very nearly in the same direction, the light becomes most intense when the difference of the routes is any multiple of a certain length, and least intense in the intermediate state of the interfering portions; and this length is different for light of different colours" (*Phil. Trans.* 1802, p. 387).

So, where will you find Young's experiment? That is the question. The answer . . .

In 1800, after having returned to London to practice medicine, Young was offered the professorship of physics at the Royal Institution, which he accepted. Two years later he was to resign on the grounds that the position interfered with his medical practice, but in 1802, before stepping down, he delivered thirty-one lectures at the institution (on Mondays, Wednesdays, and Fridays) to "apply to domestic convenience the improvements which have been made in Science, and to introduce into general practice such mechanical inventions as are of decided utility."* These popular lectures were published in 1807 as *A Course of Lectures on Natural Philosophy and the Mechanical Arts*. Those were the days. Young's lectures spanned what must have been most of scientific knowledge of the time: gravitation, Newton's laws, the solar system, animal life, vegetable life, hydraulic machines, carpentry, the nature of sound and light . . .

Toward the end of Lecture XXIII, on the theory of hydraulics, Young describes how to build a "wide and shallow vessel, with a bottom of glass, surrounded by sides inclined to the horizon" in order to exhibit conveniently many of the phenomena of waves. Today this apparatus is termed a ripple tank, and in it students create water waves and observe their behavior. Young goes on to describe the interference pattern produced by two circular waves that have been created near each other, as you might do by dropping two stones into a pond. It is the diagram of *this* interference pattern that he publishes in the *Lectures*. In other words, Young's famous illustration of interference, reproduced in textbooks worldwide, is not of light at all but of water!

To continue. In Lecture XXXIX, on the nature of light and colors, he turns his attention to interference and begins by explicitly referring to "the case of the waves of water and the pulses of sound":

> It has been shown that two equal series of waves, proceeding from centres near each other, may be seen to destroy each other's effects at certain points, and at other points to redouble them; and the beating of two sounds has been explained from a similar interference. We are now to apply the same principles to the alternate union and extinction of colours.

---

* According to Wood and Oldham, *Thomas Young, Natural Philosopher* (Cambridge, Mass.: Cambridge University Press, 1954). The published version actually contains sixty lectures!

He then does so. Because this is the crux of the matter, let us quote Young at length:

> In order that the effects of two portions of light may be thus combined, it is necessary that they be derived from the same origin, and that they arrive at the same point by different paths, in directions not much deviating from each other. This deviation may be produced in one or both of the portions by diffraction, by reflection, by refraction, or by any of these effects combined; but the simplest case appears to be, when a beam of homogeneous light falls on a screen in which there are two very small holes or slits, which may be considered as centres of divergence, from whence the light is diffracted in every direction. In this case, when the two newly formed beams are received on a surface placed so as to intercept them, their light is divided by dark stripes into portions nearly equal, but becoming wider as the surface is more remote from the apertures, so as to subtend very nearly equal angles from the apertures at all distances, and wider also in the same proportion as the apertures are closer to each other. The middle of the two portions is always light, and the bright stripes on each side are at such distances, that the light coming to them from one of the apertures, must have passed through a longer space than that which comes from the other, by an interval which is equal to the breadth of one, two, three or more of the supposed undulations, while the intervening dark spaces correspond to a difference of half a supposed undulation, of one and a half, of two and a half, or more.

Finally the climax, the wavelength of light:

> From a comparison of various experiments, it appears that the breadth of the undulations constituting the extreme red light must be supposed to be, in air, about one 36 thousandth of an inch, and those of the extreme violet about one 60 thousandth; the mean of the whole spectrum being about one 45 thousandth.

We are convinced.

Let us read it again.

Is Young describing an experiment he has actually performed? It is, upon reflection (or maybe refraction), not so clear. Young begins his exposition with an analogy to sound and water, then passes to the realm of what appears to be a concrete experiment. It is difficult to pinpoint where the analogy is to break off—indeed, whether he intends it to. What's more, the description of this experiment differs markedly in style from those of his Royal Society presentations. There he is always explicit about what he did: "I made a small hole in a window-shutter, and covered it with a piece of thick paper, which I perforated with a fine needle"; "I brought into the sunbeam a slip of card"; "I shall compare the measures deduced from several experiments of Newton, and from some of my own." And he always gives the numbers, to indecent precision.

By contrast, in the above account, he never says he did anything. It is couched in much less specific language. (This is in general true of the popular lectures, but Young usually gives credit where credit is due.) There are no details of the experimental setup (and certainly no sign of the schematic found in so many textbooks). Could it be that Young was merely performing a thought experiment or forecasting a result, in the manner of Torricelli? It seems possible, even plausible. But what about the numbers? Young does provide the wavelengths for "extreme red" and "extreme violet" light, values in good accord with modern measurements. Surely this implies some experiment. Agreed, *some*. All he says is "from a comparison of various experiments." He doesn't tell us which. And the numbers he cites are, to within rounding off, exactly those he gives in his 1801 Bakerian Lecture, obtained from Newton's own measurements on thin plates! (That is, he takes Newton's data, assumes they can be explained by waves, and gets those numbers.) My feeling is that he had those numbers on hand and used them.

I am not the only one who has noticed the ambiguities in Young's account. The historian of science John Worrall adds that Young's cavalier remark about "two very small holes or slits" is in fact fatal. The experiment is extremely difficult to perform with pinholes, and even when it is successful, the pattern on the observation screen is not one of "stripes," as Young says, but circles. To make either version of the experiment work without a laser requires, moreover, first passing the light through another hole, the collimator, to make the rays parallel—but nowhere does Young say he did this. Nor does Young ever mention his experiment again. Worrall argues that the real reason Young's theory was not accepted at the time was that no one could reproduce his results.

I do not claim that Young never performed his famous demonstration; I merely claim that the evidence is less than equivocal. To paraphrase Newton, I make no conjectures and will leave it at that. The lesson, found at every turn in the Panopticon, is that scientists have, as always, been admirably adept at reconstructing history to fit a predetermined outcome. The reasoning is almost Aristotelian, teleological—things happen for a purpose. In this case to produce the grand edifice of modern science.

Paradoxically, almost without exception do contemporary scientists reject teleological reasoning in science. It is strange they do not in history.

Young's tale, though, is not quite done. His contemporaries, including the great Laplace, dismissed Young's findings, and in the face of mounting difficulties confronted by the wave theory Young himself had second thoughts, conceding after 1810 that the wave theory could not explain certain observations. Although he continued in his medical research and to work at St. George's Hospital, he apparently did little in the way of private practice. He had too many other distractions, the most famous of which was the Rosetta Stone.

The Rosetta Stone was discovered in 1799 by Napoléon's troops in the Nile Delta. The importance of this slab of basalt, now in the British Museum, was immediately apparent to all, for the stone contained what seemed to be the same inscription in three different languages: Greek, Egyptian popular (demotic), and hieroglyphics. The stone would prove to be the key to deciphering Egyptian hieroglyphics, and today the very term "Rosetta Stone" connotes the key to cracking an unknown body of knowledge. Yet, twelve years after the Rosetta Stone's discovery, little progress had been made in deciphering it—that is, until Young, with his knowledge of Semitic languages, jumped in. Within a few months he had provided a translation of the demotic script. He then began to tackle the hieroglyphics. Jean-François Champollion (1790–1832) is generally credited with the decipherment of the Rosetta Stone's hieroglyphics, but it is a matter of record that Young made the first steps several years earlier, which Champollion acknowledged and used. Later a bitter priority dispute of international dimensions broke out; Champollion treated Young shabbily and as a result accrued all the credit. But that is, as they say, another story. Young consoled himself by contributing to the *Encyclopaedia Britannica*.

And what of the wave theory, those mounting difficulties? In this brief history a few small details have been left out, the important ones. Wrapped up in all the theories of light was the concept of the luminiferous *ether*. Sound waves propagate through air and ocean waves through water. It was natural to think that light waves required a medium in which to propagate—the ether. Everyone, including Newton, believed this. At the same time it was inconceivable that a force could be transmitted across empty space, and this too argued for the existence of an ether. But eventually, when Newton realized that his law of gravitation did *not* require any mention of such a medium in order to transmit gravity, he began to regard it as superfluous and wished to ban it from optics also. Indeed, this is one factor that led him to favor projectile theories.

Unfortunately for the wave theorists, to ascribe a set of plausible properties to the ether eventually proved impossible. If light is transmitted like sound, for instance, and we assume the ether is a hundred times less dense than air, then it should be roughly five thousand times as stiff as steel. Difficulties such as this eventually led Einstein to abandon the ether altogether. (Well, actually, see chapter 7.) Undoubtedly, the lack of a convincing picture of the ether impeded the progress of wave theories of light.

More important at the time of Young was the inability of any theory to explain certain observations connected with the polarization of light. Polarization had been causing trouble since 1669, but even more so after 1809, when Malus discovered that light reflected from a window was polarized. That is, if one looked at the reflected light through a sheet of polarizing material (the crystal tourmaline is a natural polarizer; today sunglasses will do) and rotated the polarizer ninety degrees, the light would fade out.* These results depressed Young considerably. Even worse were the experiments by François Arago in 1817 demonstrating that two polarized beams *do not interfere.*

The lack of interference between polarized beams was a severe setback for the wave theory. Today we know the reason for this is, essentially, that the two beams of polarized light are vibrating in different directions and so do not combine. Young himself made this suggestion to Arago in 1817, which amounted to a great conceptual shift. Previously, he had regarded light as analogous to sound. Sound waves are *longitudinal,* meaning that the pressure oscillates back and forth along the direction of propagation. Young realized the problem of polarization might be solved if light waves were *transverse*—like water waves, they oscillate perpendicularly to their direction of travel, and that two polarized beams were oscillating at right angles to each other.

Young's suggestion proved to be crucial in the history of optics, and with it, the wave theory began to be put on a firm basis. This is, of course, also an oversimplification. At this time the most important work was that of Augustin-Jean Fresnel (1788–1827), who a few years after Young, but independently, also developed the theory of interference of light. Fresnel's approach was far more sophisticated than Young's, and he provided a

---

* Indeed, every time you look through a pair of Polaroid sunglasses you are using Malus' discovery. Polaroids work because reflected light viewed through them loses half its intensity.

mathematically rigorous description of the entire phenomenon. Together, the work of Young, Arago, Fresnel, and others gradually forced the wave theory of light to be accepted, and the corpuscular theory was finally abandoned by the 1830s. Light had become a wave.

Until 1905. In that year Einstein used the Rosetta Stone of quantum mechanics to explain the photoelectric effect. In one stroke light became both a wave and a particle again. But that, really, is another story, which we encounter later on our tour.

3 / Joseph Henry and the (Near) Discovery of (Nearly) Everything

If America has an unsung hero, it must be Joseph Henry. While Elvis and Marilyn get postage stamps, and Benjamin Franklin announces his immortality from hundred-dollar bills, most citizens have never heard the name Joseph Henry. Yet he was the greatest American scientist after Franklin, did more to put American science on an international footing than any other person in the nineteeth century, made discoveries in electromagnetism that predated or anticipated many of his more famous contemporaries', pioneered the telegraph, and became the first secretary of the Smithsonian Institution. When one attempts to understand why this man, whose funeral in 1878 was the largest Washington had seen since Lincoln's own, has fallen into such obscurity, the reflexive answer is that scientists are not favored as American heroes. Pace. The only other suggestion that comes to mind is that Henry lacked what the public demands: glamour. His personality was uncolorful and his outer life, except for one tremendous scandal, uneventful.

Joseph Henry was a native of Albany, New York, the third child of Scottish immigrants who had arrived in the colonies in 1775. That much is known with certainty, but of his early years the rest is immediate confusion. The baptismal records of the First Presbyterian Church in Albany give his birth date as December 17, 1797, but for some reason Henry himself always insisted that the record was in error and that he was actually born in 1799. Henry obfuscated a few other things about his early life; in particular he was always reluctant to admit that his father, William, a common laborer, died of alcoholism. When Joseph was six or seven (or eight or nine, depending on whose dates you believe), his mother, Ann, sent him to the nearby village of Galway to live with her stepmother. Later in life Henry claimed that this exile took place because after the death of his father Ann could not longer cope. Now it seems that his father was still alive. The real reason Joseph was sent away probably had more to do with William's alcoholism and the birth of a new child into the family.

No records indicate that Joseph was anything but an ordinary student at Galway. Two apocryphal tales told by Henry himself survive from those

24

days and give some idea of what he considered the significant events in his life. (We remind you that *apocryphal* means "of dubious origin," not necessarily "false.") According to the first, he ordered some shoes from a cobbler but was unable to decide between square toes and round ones. Such was his indecision that the shoemaker eventually presented him with a remarkable pair—one with a square toe, the other with a round toe. Ahem. It is nevertheless true that Henry's Achilles' heel in later life would prove to be his dilatoriness. For that he would pay dearly. Even more decisive was the occasion on which he chased a rabbit under the village church only to crawl through a loose floorboard and find himself in the village library. Surreptitious visits ensued. Eventually a Mr. Broderick, the general-store manager and Joseph's employer, discovered him in his secret lair; an ardent reader himself, Broderick arranged for Joseph to be admitted to the library in more orthodox fashion.

The echoes of Ben Franklin are more than audible here, but books do seem to have made an indelible impression on the young Henry. At least one book. By 1814 or 1815 he had returned to Albany and came across an improbable tome lying on the table: George Gregory's *Popular Lectures on Experimental Philosophy, Astronomy, and Chemistry,* published in 1808. Twenty years later Henry would present the book to his own son with this inscription:

> This book although by no means a profound work has under providence exerted a remarkable influence on my life. It accidentally fell into my hands . . . and was the first book I ever read with attention.
>
> It opened me to a new world of thought and enjoyment, invested things before almost unnoticed with the highest interest fixed my mind on the study of nature and caused me to resolve at the time of reading it that I would immediately commence to devote my life to the acquisition of knowledge.

There was competition. At the time, a visit to the theater had left Henry so smitten with the stage that he even authored a play, *The Fisherman of Bughdad,* and for a few years he seriously considered an acting career. To be sure, he evinced enough talent to receive an invitation from an Albany theater with the promise of fortune and fame. At the last minute, though, he was saved from perdition by T. Romeyn Beck, the principal of the recently established Albany Academy, who offered him free tuition to study there. At least that's one version of the story. In another Henry, by then a silversmith's apprentice, did such a marvelous job at repairing a piece of scientific equipment for the academy that the folks offered him free tuition.

Well, in 1819 Henry did become an older-than-average student at the academy, which today would probably be called a prep school. From his notebooks we have a fairly accurate idea of what Henry studied: chemistry, trigonometry, geometry, English . . . the usual suspects. Navigation, calculus, surveying, natural philosophy. But remember, in those days the existence of atoms had yet to be established beyond reasonable doubt, and electricity was thought of as a "galvanic fluid."

Steam was big. Henry studied heat and steam engines; he became an assistant for one of the chemistry professors, next a tutor for the children of Stephen Van Rensselaer. When the Albany Institute, a society for the promotion of science, was organized under the presidency of Van Rensselaer, Henry became librarian. In late 1824 he read his first scientific paper before the institute, "On the Chemical and Mechanical Effects of Steam." He became a surveyor for the Great State Road project in 1825. Railways and steamboats would soon crisscross the country, and a career as a civil engineer beckoned. A professor at Albany Academy resigned during a scandal involving "a female of low character," and his position unexpectedly opened up. At the age of twenty-eight, in 1826, Joseph Henry became professor of mathematics and natural philosophy at the Albany Academy.

If the sketch seems sketchy, it is for good reason. Apart from his notebooks, which contain mostly rote exercises, we have little contemporary information about Henry's life at the academy. We know virtually nothing about the course of his intellectual development; we do not even appear to know exactly when he left off his studies at the academy (sometime in 1822). His letters, then and later, overflow with *rectitude*. So hidden are emotions under the formalities of the times that a reader of our era has the impression he is eavesdropping on characters from *Sense and Sensibility.* "When I visited [Niagara] falls in Oct. 1825," Henry writes during one of his surveying expeditions, "they did not answer my expectations. Perhaps I did not view them with the propper [*sic*] poetical feelings, but in this visit I have been more fortunate." Henry's inaugural address as a professor at the academy in September 1826 gives a further impression of his style. After a long summary of the history of mathematics he concludes with a paean to the virtues of studying the queen of sciences. Henry's spotty education comes through in the spelling:

> Pure mathematics requires no experimental illustration and from the effects it produces on the reasoning faculties is rendered more emenently useful as a branch of liberal education. The powers of the mind are as

much invigorated by frequent exertion as the limbs are strengthened by repeted action. . . . Pure mathematics above all other studies serve to call forth this spirit of intellectual exertion. By an early attention to this science the student acquires a habit for of reasoning & an elevation of thought which fixes his mind & prepares it for every other puruit. And although in the active and more important duties of afterlife he may have forgotten every proposition in geometry & every principle in algebra still he will be much indebted to those early studies for that general discipline & enlargement of the understanding so necessary to his professional rise & usefulness.

It had been said by Franklin, and it is said by every mathematics professor today. As Plato might have put it, "All knowledge is repetition." Yet Henry was no mathematician. One would never guess from his remarks—or anything else known about him—that over the next five years he would make some of the most fundamental discoveries of the century in electromagnetism.

When Henry took up his teaching post at the academy, the science of electromagnetism was precisely six years old. Electricity and magnetism as distinct subjects of investigation sent their roots into the mists of antiquity. The Greek word for "amber" is *electron,* because the ancient Greeks had been aware that rubbing amber with fur produced a static electric charge. Ancient Romans understood that rubbing iron needles with lodestone caused them to attract each other, a property so astounding that they regarded it as supernatural. By the opening of the nineteenth century, batteries had been invented, as had telegraphs that ran on them (chapter 11). Natural philosophers had long held suspicions that electricity and magnetism were somehow related, but all attempts to establish the connection failed.

Until 1819. In the winter of 1819–1820 a professor in Copenhagen, Hans Oersted, noticed that a current-carrying wire would deflect a compass needle. This accidental discovery was the first in the history of electromagnetism and probably the greatest, for it established the long-sought link between electricity and magnetism. From then on the two forces would be regarded as aspects of some single, underlying phenomenon, yet to be elucidated. Discoveries came fast and furious after the publication of Oersted's pamphlet in July 1820. Within two months André-Marie Ampère had discovered the laws relating electrical current to magnetism that bear his name. During the same weeks Johann Schweigger in Prussia found that the deflection of Oersted's compass needle could be doubled by making the wire run in both directions. By wrapping a coil of wire around the compass, the deflection could be made arbitrarily large. The "multiplier," as Schweigger called it, would prove the basis for all future electrical meters, the electromagnet and the telegraph. The electromagnet itself came into existence in 1825, when William Sturgeon of England bent a piece of soft iron into the shape of a horseshoe, wrapped a few loose turns of wire around it, sent a current through the thing—and found it could lift nine pounds.

That's as it's usually told. On the other hand, for the past 150 years claims have periodically surfaced that one Italian, Gian Domenico Romagnosi, announced Oersted's result in 1802, nearly two decades before the Dane. If true, every physics textbook in existence would need to be rewritten. It can't be denied that a report of Romagnosi's experiment did appear in an Italian newspaper in 1802. Thomas Coulson, one of Henry's biographers, accepts Romagnosi's priority for the discovery. But in reading the newspaper's description of the demonstration, I must agree with John Fahie in *A History of Electric Telegraphy* that if any effect was observed, it must have been a mistake, as the experimental setup doesn't appear to have even included a closed circuit, which is a well-known requirement to produce electric currents.* Nevertheless, a street is named after him in Florence, Italy.

In any case, Romagnosi's experiment was forgotten; to the world, and Henry, electromagnetism was born in Copenhagen. Henry's own researches into the field began rather casually. On September 21, 1827, he wrote to Lewis Beck that "as usual I am busy engaged in applying *birch* to one end and *arithmetic* to the other," and then goes on to describe philosophical activities. In particular, he reports a grand exhibition of the northern lights. Popular explanations were that it was a lunar rainbow, a reflection from a cloud, or "a sign in the heavens that Mrs. Whipple should have been hung with Jessy Strang." Henry retired to the laboratory. An assiduous reader of European journals, he had studied an anonymous 1821 article by Michael Faraday on the history of electromagnetism in which an experiment of Ampère was described. Ampère believed that terrestrial magnetism—including the aurora borealis—was a result of electrical currents, and to demonstrate it, he connected a powerful battery to a semicircular loop of wire that was free to rotate. Henry merely reproduced the

---

* A needle connected by wire to one terminal of a battery (a voltaic pile) was brought near a compass needle, and the compass needle was deflected. However, the other terminal of the battery does not appear to have been connected to anything! It is easy to imagine that some static buildup on the probe caused the deflection. I have attempted to duplicate Romagnosi's results without success, but then again I am a theorist, not an experimentalist.

experiment. When a current of "galvanic fluid" was passed through the loop, it "placed itself in a plane perpendicular to the magnetic meridean [east-west] and thus in mineature, as it appears to me, represents the auroral arch." Essentially Henry, following Ampère, had constructed an electrocompass.

From this modest start, Henry's investigations rapidly progressed. He had intended to perform galvanic experiments on the body of the hanged murderer Strang—conceivably with a view to revive the dead—but was apparently frustrated. Nevertheless, three weeks after his letter to Beck he presented his first paper on electromagnetism before the Albany Institute, in which he described how Schweigger's multiplier could be applied to enhance a variety of electromagnetic demonstration devices. Over the next eighteen months, he adapted the coil to a variety of circumstances, in particular to the electromagnet.

Sturgeon's original magnets were feeble affairs, loosely wrapped with a few turns of uninsulated wire. Teaching himself how to insulate the wire with silk or shellac and wrap it tightly around the horseshoe, Henry was soon constructing vastly more powerful devices. One of these was a magnet capable of lifting 650 pounds, "probably the most powerful magnet ever constructed." Unfortunately, he wasn't moved to publish his results until he "had the mortification" of learning that Gerrit Moll of Utrecht had created Europe's largest magnet, capable of lifting 150 pounds. In what would soon prove to be a typical and detrimental pattern, Henry quickly wrote up his experiments of the previous several years and persuaded Yale professor Benjamin Silliman, who published the country's first scientific journal, *American Journal of Science and Arts,* to rush an article into print. Henry's paper appeared in the January 1831 issue after a reprint of Moll's with this too accurate disclaimer: "The only effect Prof. Mohl's [*sic*] paper has had over these investigations is to hasten their publication." A few months later Henry provided Silliman with a magnet capable of hoisting an unprecedented 2,063 pounds.

Electromagnets and procrastination would prove central to virtually all of Henry's work. The result was a series of great discoveries and priority disputes. Of the discoveries, the relay, developed over the next years, proved to be the basis of the electromagnetic telegraph, a version of which he demonstrated to his students in 1831. Later that year Henry also unveiled a little device that consisted of an electromagnet in the form of a seesaw that would rock back and forth seventy-five times a minute. This "philosophical toy" appears to have been the first electromagnetic motor. Two years later, William Ritchie in England produced something similar, and Henry quipped that "Mr. Richee [*sic*] has lately reinvented my machine." His annoyance at Ritchie continued at least two decades: at one point Henry even wrote to a French editor, asking him to retract an attribution to Ritchie of the first electromagnetic machine.

Henry's research at the time into "pure" science ultimately had greater technological repercussions than his technological innovations. We speak of induction, the basis of modern civilization. In August 1831, Michael Faraday (1791–1863), often considered England's greatest experimental scientist, was carrying out investigations remarkably similar to Henry's (indeed, their entire careers seem to be a case of parallel lives). Oersted had shown that electricity could produce magnetism. Faraday, consciously under the spell of Newton's third law—"for every action there is an equal and opposite reaction"—held "an opinion almost amounting to a conviction" that magnetism should be able to produce electricity.

In the course of his experiments, Faraday wrapped several separate coils of wire around a ring of soft iron. To one coil—the "primary"—he could connect a battery; to the "secondary" coil he attached some external wires that ran over a compass needle. When Faraday connected the battery to the primary, he noticed a small, momentary flicker of the needle, about three feet away. The electric current in the primary had somehow—across space—*induced* a current in the secondary, which in turn produced a magnetic field that acted on the needle. Faraday had invented the transformer, the basis of the entire electrical power industry and a crucial component of virtually every electrical appliance besides. In October, Faraday devised the decisive test: He thrust a bar magnet through a coil of wire and observed the deflection of a galvanometer needle. Magnetism could indeed produce electricity. A month later, Faraday announced his findings before the Royal Society.

Henry was not amused. Although Faraday was using Henry's magnet design in this experiments, he did not explicitly acknowledge the American (probably he did not feel this was central to the experiment). More to the point, Henry claimed to have performed the same experiments before his competitor. Unfortunately, he failed to publish and so perished. To be precise, in August 1831 he had begun a series of investigations that he was forced to interrupt. What exactly he did at that time we simply don't know. A preliminary notice of Faraday's work was published in December 1831, but Henry probably didn't receive it until the following spring. The vague report was enough to galvanize him into continuing his interrupted experiments. A fuller report of Faraday's work appeared in the *Philosophical Magazine* of April 1832, which also probably got to Henry a few months late. A description of Faraday's transformer experiment was still lacking, but the article spurred Henry on to even greater frenzy. Again he quickly wrote up his results for Silliman and sent them off.

What Henry claims he did "before having any knowledge of the method given in the above account" was to wrap an "armature" (an iron bar) with wire and place it across the ends of one of his big horseshoe electromagnets. Thus he created a device almost identical to the one Faraday would build or was building across the Atlantic. The results were identical too: the momentary deflection of a compass needle about forty feet away. Henry had created (maybe) the first iron-core transformer and with it illustrated "most strikingly the reciprocal action of the principles of electricity and magnetism, if indeed it does not establish their absolute identity."

Of course, Henry penned these words only after learning of Faraday's results. He grumbled, ironically, that "no detail is given of these experiments," but to a first-rate scientist the result is the most important thing to know; how to get there often becomes evident. The discovery of induction was one of the supremely great scientific events of the nineteenth century. But although Americans often give Henry priority—some even claim without documentation that he found it as early as 1829—in light of his behavior I think it proper that to Faraday goes the credit. Henry himself never contested Faraday's priority. He understood that if you snooze, you lose.

Nevertheless, during those researches Henry did notice that when a battery was *disconnected* from a wire thirty or forty feet long, a spark would appear between the wire and the battery terminal. This was Henry's discovery of "self-induction." When a current in a wire collapses, it produces a magnetic field, which in turn induces a momentary reverse current in the wire. Self-induction is generally regarded as Henry's greatest scientific discovery; no one before him had even suspected its existence. Three years later Faraday would independently discover it for himself.

Induction was far from the end of the story. In 1832 Henry left Albany for a professorship at Princeton University, then the College of New Jersey, where he continued his teaching and researches. (To this day you can view many of Henry's instruments in the Physics Department.) He gradually got better about publishing his results on time, but he had been stung once too often and became inordinately sensitive about receiving due credit; more than once he engaged in writing anonymous articles in which he assigned himself a prominent place in the history of electromagnetism.

All the while his reputation was growing on both sides of the Atlantic, and when in 1837 he was able to take a leave of absence, he traveled to Europe, befriending Charles Wheatstone and Faraday. Sadly, Henry's diaries are as impersonal as his letters, and we do not know much about how he got on with his great and worthy rival. It appears to have been well. To Wheatstone he explained how to overcome the technical hurdles the latter was experiencing with his nascent telegraph; he bested both Englishmen in an experimental contest involving induction. On his return to the States, Henry became acquainted with one Samuel F. B. Morse, who was also developing a telegraph and also experiencing technical difficulties. Rather than patent his own devices, over the next six years Henry advised Morse on the construction of magnets and relays. The result: one of the great technological scandals of the nineteenth century. That story is recounted in excruciating detail in chapter 11.

Henry's experiments at Princeton were many and varied. He continued to parallel Faraday; he constructed an artificial electric eel to understand why the animals did not electrocute themselves; he further developed the transformer. In 1845 he became the first person to measure the temperature of sunspots. Perhaps his most extraordinary experiments were those of 1842, by which he established that powerful sparks given off by Leyden jars (primitive capacitors) and "thrown onto the end" of a wire circuit in an attic were capable of magnetizing needles across space in "a parallel circuit" in the basement. Fifty years later Heinrich Hertz (chapter 6) would perform essentially the same experiment and be immortalized as the discoverer of radio waves.

Did Henry in fact discover radio waves a full half century before Hertz? Apparently. (There have been disputes.) Whatever the case, Henry performed his experiments two decades before Maxwell had developed his theory of electromagnetism, and without the theoretical context he couldn't have understood what he was observing. In this Henry was in much the same postion as Niepce de Saint-Victor upon discovering radioactivity in 1857 (chapter 5) and Penzias and Wilson upon discovering the cosmic microwave background radiation in 1965. Nevertheless, Henry had ideas, prescient ideas. "After a laborious investigation of the phenomena we have described," he wrote in an anonymous *Encyclopaedia Americana* article, "Professor Henry has succeeded in referring them all to the hypothesis of the existence of an electrical plenum, through which dynamic induction is transmitted wave fashion; and showing that, in all cases of the disturbance of the equilibrium, the fluid comes to rest by a series of oscillations." Elsewhere he mentions that the "diffusion of motion . . . is almost comparable with that of a flint and steel in the case of light." Maxwell credited Faraday with having proposed the electromagnetic theory of light in 1846, but in this case it is clear Henry got there first. What it needed was precisely Maxwell to show that electromagnetic waves and light were indeed one and the same.

By 1846 Henry was certainly the foremost scientist in America, but he was dissatisfied with the recognition he had received. Others were making money off his innovations; Princeton was paying him a lousy salary. Faraday had recently scooped him again with a discovery, now famous, connecting magnetism and polarized light. Clearly he was looking for a change and flirted with offers from Harvard and the University of Pennsylvania. What he did instead was help found the Smithsonian Institution and, as no good deed goes unpunished, was elected its first secretary. Henry was largely responsible for building the Smithsonian into the institution we know today. He had his share of triumphs and scandals there, involving Stephen A. Douglas for one, but because these are more political than scientific, we leave them aside. The Smithsonian was not enough to keep him busy, though. Henry became the chief organizer of science in America, one of the founders of the American Association for the Advancement of Science and the second president of the National Academy of Sciences, a post he held eleven years. Not to mention the American Association for the Advancement of Education, the Washington Philosophical Society, the Permanent National Commission during the Civil War, the Light House Board . . . Science did not completely fall to the side either. Among other things, Henry found time to give advice to Alexander Graham Bell regarding an invention that we will pursue again through a labyrinth of infinite convolutions.

Henry's reputation is a strange one. By the time he died in 1878, he was hailed as "the Nestor of American science," and as I have said, his funeral was the largest Washington had seen since Lincoln's own. His work was internationally recognized by his colleagues. In 1893 the International Congress of Electricians unanimously voted to make the "henry" the standard unit of inductance. Yet despite the statue in his honor that stands before the Smithsonian and his central role in this country's scientific history, most Americans have no idea of who he is.

One of the reasons for Henry's eclipse is surely nationalism. Although it was foreigners who nominated his name for the unit of inductance, in Europe to this day he is regarded as Faraday's shadow. In *A Short History of Technology*, written by two Englishman, Faraday gets at least a dozen mentions, Henry none. And since American scientists in the twentieth century looked toward Europe for their education, they have also unconsciously bought the European slant.

Henry's personality does not help his cause. Franklin wrote an autobiography, penned memorable remarks, and had the advantage of being a politician besides. Henry, on the other hand, is a puzzle. Letters and diaries exist. But regardless of how many of these you read, they remain highly impersonal, and you are left constructing his character through fragments. Nevertheless, Henry deserves a visit in the Panopticon, not so much because he is forgotten but to remember how much we have forgotten.

Surely he deserves a postage stamp.

4 / Neptune: The Greatest Triumph in the History of Astronomy, or the Greatest Fluke?

For sheer color and uproar, few tales of the Panopticon can match that of the discovery of Neptune. The events, when they first took place, practically caused a breach of diplomatic relations between France and England, and even today, one and a half centuries later, partisans can be heard echoing the original cries across the depths of cyberspace and from the pages of encyclopedias.

The story is one of the most famous in astronomy. As you read it in the fourth edition of William Kaufmann's *Universe,* the most widely used introductory astronomy text, it goes (roughly) like this: On March 13, 1781, the oboist and astronomer William Herschel sighted an object in the sky that he took to be a new comet. Repeated observations by himself and others soon confirmed that this was no comet but an entirely new planet. Herschel proposed that it be known as Georgium Sidus, or the "Georgian Star," after King George III, but rather more sensibly the Prussian astronomer Johann Bode hit on the name Uranus. Which stuck. Uranus, as it turned out, had been sighted frequently by astronomers before Herschel, having been invariably mistaken for a star, but the old and new observations together soon led astronomers to an incontrovertible conclusion: Uranus' orbit was not behaving as Newton's laws would have remotely predicted. The astronomer royal wondered aloud whether Newton's law of gravity should be repealed.

In 1843 a young student at Cambridge, John Couch Adams, began to calculate the perturbations to Uranus' motion that would be caused if a new, undiscovered planet lay beyond Uranus' orbit. After two years of hard work, he could *predict* where this new body should lie. Adams reported his results to the astronomer royal, George Airy, who ignored them. Unfortunately for Adams and England, within a few months a Frenchman, Urbain Le Verrier, published a similar series of calculations, which agreed remarkably well with Adams' own. Shocked into action, Airy put his staff on the hunt for this planet, which turned up negative. But Le Verrier had meanwhile written to Johann Gottfried Galle, a Berlin astronomer. The letter arrived on September 23, 1846. That very night Galle and an assistant pointed their telescope at the region Le Verrier had indicated, and . . . Neptune.

War ensued.

The same story can be found in Eric Chaisson and Steve McMillan's textbook *Astronomy Today,* and doubtless in countless other places. The *New Catholic Encyclopedia* concludes, "In this way Le Verrier gave the most striking confirmation of the theory of gravitation propounded by Newton." *Britannica* itself seconds this judgment: "The visual discovery of Neptune in just the position predicted constituted an immediately engaging and widely understood confirmation of Newtonian theory." Neptune's discovery was, in fact, considered history's greatest confirmation of Newtonian physics, a claim found then and now everywhere; after Neptune, nothing was left to explain.* The *New Catholic Encylopedia,* however, should explain one significant omission: Adams. There is no mention of him. Rather, we are treated to a page on Le Verrier's religious convictions. Neither does the *Larousse Encyclopedic Dictionary* mention the Englishman, though he does get one line in the full *Larousse Encyclopedia.* Actually, no explanation here is needed. In the Panopticon, nationalism lives.

What is really astounding about all these accounts is the other omission: that the whole thing was a fantastic coincidence.

"The strange series of wonderful occurrences of which I am about to speak," wrote Benjamin Apthorp Gould in his splendid 1850 report to the Smithsonian on the discovery of Neptune, "is utterly unparalleled in the whole history of science."

He may have been correct. In 1841 John Couch Adams, then a young student at Cambridge University, ran across a book by the astronomer royal, George Biddell Airy, in which Airy discussed the problems with Uranus. "With respect to this planet," Airy conceded, "a singular difficulty occurs." Basically, regardless of what selection of data astronomers used to predict the path Uranus should follow according to Newtonian mechanics, within a few years telescopic observations showed that the planet had drifted significantly from the predictions. This had been going on for sixty years, since William Herschel's original discovery. Physics was in crisis. Some natural philosophers thought Uranus might have been hit by a comet, which perturbed its orbit. Others speculated that the seventh planet might be moving through a resistive medium, which would cause

---

* Even the invariably sober Steven Weinberg repeats this in his graduate-level general relativity text, *Gravitation and Cosmology* (New York: John Wiley and Sons, 1972).

the orbit to slowly spiral inward. Still others thought that Newton's law of gravity might not be valid at such large distances from the sun. A few thought that Uranus' meanderings might be due to the gravitational tug of an unknown planet.

Within a few days of having purchased the book, Adams had decided to tackle the problem of Uranus from the hypothesis that a new, undiscovered body caused Uranus' errant peregrinations. This required calculations, horrible calculations, by hand. But the young scholar was up to the task. Born in 1819, Adams was something of a mathematical prodigy. By age sixteen he sang, played violin, calculated orbits, and predicted solar eclipses. At Cambridge, in 1843, he would become senior wrangler in the Mathematical Tripos, outperforming his nearest challenger on this legendarily difficult exam by two to one.* Soon afterward, he received the first Smith's Prize, the most prestigious mathematical trophy the university had to offer.

Having put the Tripos behind him, Adams could devote his entire attention to the New Planet, and he spent much of the next several years immersed in computations, which they say filled ten thousand pages. Basically, the strategy was to imagine a new planet with a given mass and distance from the Sun and to calculate its gravitational effect on Uranus. But this sort of calculation—known as a perturbation calculation—is at best tedious, at worst fatal. Before he even began, Adams had to check extant astronomical tables and even rederive them, a miserable project in itself. Nevertheless, will triumphed over algebra. By autumn 1845 Adams had his answer: The position of the New Planet was in hand.

That was the easy part. The hard part was to convince the powers that be to search for the New Planet. In 1845 all the planets had been known since antiquity, with the exception of Uranus, and that one had been discovered by accident. Experimentalists are a conservative lot and ever suspicious of theoretical types; the idea of looking for a planet based on some numbers sat with astronomers about as well as the idea of hunting for quarks did with physicists in the 1950s. The situation was made

---

* There is a story told at both Cambridge and Oxford that an American visitor, rummaging through the rules and regulations of the university from the middle period (that is, from about the fifteenth century), discovers that a student has the right while taking the Tripos to demand a pint of ale. During the Tripos, the visitor, dressed in subfusk for the occasion, forthrightly rises and demands his pint. The envigilator, without batting an eye, leaves the room, crosses the street, and a few moments later returns from the nearest pub with a pint of ale in hand. The exam continues. A few days later the visitor is summoned before the University Judiciary Commission and fined for not wearing a sword to the examination.

worse by the fact that Adams was by nature a shy and diffident character. Did he attempt to publish his work? No. Rather, with a letter of introduction in hand, Adams traveled to the Royal Greenwich Observatory to present his results in person to the astronomer royal. But—he had made no appointment. Airy was in France. Adams left behind the letter of introduction from James Challis, director of the Cambridge Observatory, to whom he had also entrusted his calculations. Upon returning from abroad, Airy replied to Challis, inviting Adams to summarize his results in a letter.

Rather than write the letter, Adams decided to present his conclusions in person. Once again he drops by the Royal Greenwich Observatory unannounced. Airy is out. Adams replies he will come back in an hour. He strolls around the garden, returns. Airy is having dinner and cannot be disturbed. Adams leaves the summary of his calculations with the butler and returns to Cambridge.

Airy ignores Adams' results.

The plot thickens. Airy did not ignore Adams' results. Being an observationalist, he was genetically conservative and skeptical. He was also busy. His wife was about to give birth to their ninth child; an assistant at the observatory had been charged with murdering an infant he had fathered upon his own daughter; he had an observatory to run. He was also busy filing. Airy, by all accounts, was a strange bird. He evidently never threw out anything (one reason we know so much about the case), even endorsing pieces of blotting paper he had wiped his pen on, filing them with the rest. But he did answer Adams, raising some objections and asking for more details. Adams, however, never replied. He was busy refining his calculations.

Urbain Le Verrier was also busy. Le Verrier, born in 1811, was, like Adams, a gifted mathematician who had graduated from L'Ecole Polytechnique, the glory of French science. In 1837 Le Verrier, initially trained as a chemist, was offered a post in astronomy at the Polytechnique. He accepted. "I have already begun to mount the ladder—why should I not continue to climb?" Le Verrier's vanity and ambition were as legendary as Adams' self-effacement and abhorrence of the limelight. Colleagues referred to Le Verrier as a *mauvais coucheur* (a "bad bedfellow"), an epithet, I assure you, deserved by more than one scientist in history. And he proved to be a great mathematical astronomer. Since 1840 Le Verrier had been working on planetary orbits, and in the late summer of 1845, when François Arago, director of the Paris Observatory, suggested that he turn his attention to the problem of Uranus, Le Verrier jumped at the chance. Within a few months he presented his first paper to the French Academy of Sciences.

The paper, which showed that perturbations from Jupiter and Saturn were insufficient to account for the motion of Uranus, found its way across the Channel to Greenwich by the end of the year, but it did nothing to persuade Airy to take Adams' work more seriously. Six months later, on June 1, 1846, Le Verrier presented his next paper on Uranus. This one contained (almost) everything. Working by somewhat different methods than Adams, he concluded, like his unknown rival, that the New Planet could be responsible for the anomalies in Uranus' orbit. Like Adams, he predicted where the planet would be. And like Adams, he narrowed the location to an area between the constellations Aquarius and Capricorn. Unlike Adams, he published his results.

When Airy received Le Verrier's second paper he wrote back to the Frenchman with further queries and objections, but Le Verrier answered all of them. Given that Le Verrier's calculations so closely agreed with Adams', the astronomer royal was now convinced that an undiscovered world lay waiting to be possessed. Then and there he authorized James Challis at Cambridge to begin searching for the New Planet with one of the largest telescopes in England. Inexplicably, Airy not only failed to mention Adams' calculations to Le Verrier—indeed, keeping them secret from nearly everyone except Challis, who had received them directly from Adams— but also never bothered to inform Le Verrier of Challis' search. One can only speculate that Airy, a Cambridge graduate, wanted the double honor— prediction and discovery—to go to Cambridge and England should the search succeed, but the ridicule to go to the Frenchman should it fail.

The search failed. On July 29, 1846, Challis pointed his telescope to the region indicated by the theoretical calculations and began comparing star positions night after night to determine whether anything had moved. The sweeps went on throughout August. Nothing.

Meanwhile, across the Channel, newspapers throughout Europe were beginning to pick up the story, confidently asserting that Le Verrier's planet would soon be discovered. "Analysis transports us to the regions of the unknown," opined the French newspaper *Le Constitutionnel*, "and brings us back laden with the most splendid discovery. . . . Let us hope . . . we shall soon succeed in seeing the planet whose position has been ascertained by M. Le Verrier."

Media hype and scientific progress have always been two different things. Le Verrier was actually having his own troubles convincing the authorities to mount a serious search for the planet. As we have said, experimentalists are a conservative lot. Gould pointed out in his report to the Smithsonian, "It must, indeed, be confessed that astronomers in general did not seem to consider the theoretical results, published by Mr. Le Verrier, as necessarily indicating the *physical* existence and true position of a such an exterior planet." Or as Challis himself put it, it was "so novel a thing to undertake observations in reliance upon merely theoretical deductions; and that, while much labor was certain, success appeared very doubtful." That was the first problem. The other difficulty was that in his second paper Le Verrier had not located the New Planet's position with pinpoint accuracy, but only to within ten degrees. This is not an insignificant swath of sky, being about twenty times the diameter of the moon. Like Adams before him, Le Verrier had already gone back to refine his calculations.

Despite the psychological and astronomical obstacles, desultory searches were undertaken in France and England, and one was proposed in America. The English search took place at the Greenwich Observatory itself. Inexplicably, neither Airy nor Challis informed the astronomer in charge of Adams' results, and the hapless fellow had to rely solely on Le Verrier's. But then, as now, telescope time was at a premium, and when after a few weeks the French and English searches turned up nothing, they were abandoned, just as Challis was abandoning his. The American search never got onto the telescope schedule.

On August 31 Le Verrier published his refined calculations. Two days later Adams finished his own and finally wrote to Airy—after a year. Airy is on vacation. This time Adams prepares a paper to present at the annual meeting of the British Association for the Advancement of Science. He arrives too late for his session and fails to deliver his talk. Meanwhile, Le Verrier is campaigning across Europe to initiate a search for the New Planet. He canvasses French, German, and Russian astronomers without success. Finally he recalls a dissertation he received a year earlier from Johann Gottfried Galle, an assistant at the Berlin Observatory. Le Verrier writes Galle, including a few lines of flatttery as a recompense for the year's delay, then tells him to get on with the search. Galle obliges. He pesters the director to allow him use of the telescope for the evening; the director yields. A student, Heinrich d'Arrest, begs to take part. That very night, September 23, the hunt begins. D'Arrest recalls that the observatory owns the Berlin Academy's star atlas, the most modern and detailed star atlas known to man. They carry it with them into the dome. One after another, Galle calls out the coordinates of the stars within the telescope's field, and d'Arrest checks their position against the atlas. Galle calls out a particular faint star and its coordinates. Silence. A perplexed silence. Finally d'Arrest exclaims, "That star is not on the map!"

Within days, hours, Le Verrier is Europe's hero, the man who predicted the existence of a new world. Newspapers toast him, the academy meetings overflow when he is rumored to be in attendance, the king requests an audience. Le Verrier takes his fame as only deserved. Galle, who first observed the new planet, suggests the name Janus. Le Verrier objects on the grounds that Janus would imply that the new planet is the last in the solar system, and counters, falsely, that the Bureau of Longitude has chosen Neptune. But as his celebrity mounts, he changes his mind. The new planet should be called Le Verrier. François Arago obliges and announces the decision to the academy.

Airy is still on vacation. John Herschel, William's son, decides that there is no time to lose in staking England's claim to the discovery. He writes to the London paper *The Athenaeum* and for the first time discloses Adams' role in the affair. When Airy returns he realizes he has a lot of explaining to do. To make matters worse, when Challis looks over his data, he realizes he had captured Neptune not just once but twice in the first days of his search and failed to recognize it. Airy attempts to portray himself as an impartial arbiter, one who wanted to be certain Adams' calculations were corroborated by Le Verrier. But when Le Verrier receives Airy's letter, in which it becomes clear that the astronomer royal knew of both calculations all along, he becomes furious.

National pride is at stake. The great Arago again mounts the podium at the academy. "Le Verrier is called upon today to share the glory, so loyally, so rightly earned with a young man who has communicated nothing to the public and whose calculations more or less incomplete, are totally unknown in the observatories of Europe! No! No! The friends of science will not allow such a crying injustice to be perpetrated!" The French papers go ballistic, calling the English action "a planetary theft." Cartoonists have a field day. Airy continues to justify his actions. Challis—who had Adams' calculations in hand a full year without acting, and who when finally ordered to action bungled the job—loses all credibility. The bitter controversy rages on for years, decades, centuries . . .

*   *   *

The end (if one can call it that) was as surprising as the rest, in every respect. By 1847 an American astronomer, Sears Cook Walker, and the German Adolf Cornelius Petersen independently discovered that a Frenchman, Michel Lalande, had observed Neptune in 1795, noted several days later that it had moved, and assumed that his first observation was erroneous. In any case, by Walker's and Petersen's parallel detective work, "it was thus rendered a moral certainty" that Neptune had been spotted fifty years earlier, and this extra data point allowed a more accurate determination of the new planet's orbit. Furthermore, on October 10, 1846, only a few weeks after Neptune's discovery, an English amateur, William Lassell, discovered a moon of Neptune—Triton. By observing the period of the moon's orbit, an accurate mass for Neptune itself could be computed via Kepler's third law, a method every physics student learns in the first semester.

You see, the calculations of Adams and Le Verrier were not assumption-free. Both men needed to guess the mass of the unknown planet as well as its distance from the Sun. The new data showed that Neptune was not nearly as far from the Sun as Le Verrier and Adams had supposed, and that the mass was only about half as great as they had assumed.* With the revised orbit and mass in hand, Professor Benjamin Peirce (1809–1880) of Harvard, considered the greatest American mathematician of his day, subjected all the calculations to reanalysis. At the March 16, 1847, session of the American Academy of Arts and Sciences in Boston, Peirce made an extraordinary announcement. "From these data," the secretary communicated,

> without any hypothesis in regard to the character of the orbit, he has arrived at the conclusion that THE PLANET NEPTUNE IS NOT THE PLANET TO WHICH GEOMETRICAL ANALYSIS HAD DIRECTED THE TELESCOPE; that its orbit is not contained within the limits of space which have been explored by geometers searching for the disturbances of Uranus; and that its discovery by Galle must be regarded as a happy accident.

Apart from being able to calculate a better orbit for Neptune with the new data, Peirce also pointed out a subtle flaw in Le Verrier's method. Le Verrier had assumed the New Planet's distance from the Sun lay within

---

* They assumed that Neptune's distance was between thirty-five and thirty-eight astronomical units (between thirty-five and thirty-eight times the Earth's own distance to the Sun) when in fact Neptune's actual distance is thirty astronomical units.

certain bounds, and he applied his method uniformly across this range. It turns out, though, that within this range lies a critical distance at which the period of Neptune's orbit (its year) is exactly two and one-half times Uranus' year. When planetary years occur in such simple multiples, strange resonances occur, causing very large and irregular perturbations in the orbit. The bottom line: One cannot use Le Verrier's method in this vicinity.

But if Peirce was correct, how on earth did Galle stumble across Neptune?

It was a fluke. In assigning Neptune a distance that was too great, Adams and Le Verrier underestimated the planet's gravitational influence on Uranus, but they compensated by making Neptune's mass too great as well. Thus they forced the gravitational perturbations on Uranus to be correct even with the wrong orbit. Yet during the epoch around 1800, it turned out, the actual orbit of Neptune lay very near those computed by the two mathematicians. If the search had been carried out even five years later, the planet would not have been found.*

Peirce's conclusion, "as paradoxical as it might at first have appeared to many, was announced with a candor and moral courage only equalled by that of Le Verrier in his original prediction of the planet's place." Candor and moral courage are rarely universally appreciated. Upon publication of Peirce's analysis, a scientific controversy as intense as the political one erupted. It was not helped by the fact that Peirce himself retracted a few of the particulars. American and European scholars arrayed themselves into phalanxes, taking transoceanic positions for or against the Harvard professor. America's premier orator and president of Harvard, Edward Everett, joined the fray by addressing the American Academy of Arts and Sciences, imploring its members "not to sanction so improbable a declaration." Joseph Henry himself sighed that Peirce had been "too hasty" in publishing his results. He never, however, accused Peirce of being wrong. To the contrary, the first secretary of the newly formed Smithsonian Institution paved the way for Gould's report on the matter.

---

* One might well ask how on earth both men could have picked the same erroneous figures for the distance to the unknown planet. They were influenced by Bode's law, a famous heuristic rule of the day that supposedly gave the distances to all possible planets, but which is known to be nonsensical. Once they chose the distance, the mass of Neptune followed in order to get the size of the perturbations correct. Regarding the legendary difficulty of the calculations, the famous mathematician John E. Littlewood in his book *A Mathematician's Miscellany* (London: Methuen, 1953) shows how to do it in about five pages. His remarks also lead one to believe a large amount of luck was involved. He regards the discovery of Pluto by similar methods as "a complete fluke."

Le Verrier, predictably, went nonlinear, publishing several articles in the *Comptes Rendus* in which he attempted to refute the American. He claimed that the orbit computed by Walker on the basis of the 1895 data was inaccurate. He claimed that his own calculations could not be expected to have an error of less than 10 percent. He claimed that Peirce's mass, which was one-half what he himself had used, would only make an error of about 20 percent in the diameter of Neptune. He claimed that the new mass of Neptune, deduced by Kepler's third law, was inaccurate.

Walker's orbit inaccurate? Not compared to the previous one. Ten percent? Ten percent is not at issue. The wholesale discrepancies between the computed and actual orbits of Neptune are. The diameter of the New Planet? Totally irrelevant. Determining mass by Kepler's third law inaccurate? Perhaps but not compared to guessing.

John Herschel also waded into battle on the side of Le Verrier and Adams. His arguments, for this reader at least, are difficult to understand. Herschel fully accepted the results of Walker's and Peirce's calculations, that the orbits of the hypothetical Neptune and the real ones differed considerably, yet he refused to find any accident in the eighth planet's discovery. "Posterity will hardly credit," he wrote, "that . . . not only have doubts been expressed as to the validity of the calculations of those geometers and the legitimacy of their conclusions, but those doubts have been carried so far as to lead the objectors to attribute the acknowledged fact of a planet previously unknown occupying that precise place in the heavens at at that precise time, to sheer accident!" Indeed, after Peirce had retracted one of his objections, Herschel declared that the issue of Neptune's identity was now "defunct"—the theoretical and observed planets were the same.

It seems to me that the opinions of these worthy ancestors are moot. With the help of modern computers one can retrodict the orbit of Neptune to any date desired, and there is no question that the real Neptune does not follow the trajectory predicted by Adams and Le Verrier. Peirce was correct.

There is a further denouement to the tale. Adams (who, by the way, was the only actor in the drama to remain silent) and Le Verrier met at a conference in Oxford in 1847, while the controversy was still raging. Against all odds, the two men became friends for the rest of their lives. John Herschel had tried to get both of them awarded the Royal Astronomical Society's Gold Medal, but splitting the honor was against the rules, and neither candidate alone received enough votes; the medal was awarded to neither. By 1848 Herschel managed to have them both presented a special "testimonial," and with this judicious peacemaking the saga of Neptune was to the public eye largely laid to rest.

Almost. Queen Victoria eventually offered Adams a knighthood, but he decided he couldn't afford it and declined. John Adams died in 1892. Three years later a memorial plaque was placed in his honor in Westminster Abbey, near the graves of Isaac Newton and Charles Darwin. George Airy died within three days of Adams, but his reputation had been so tarnished by the Neptune affair fifty years earlier that no memorial at Westminster was erected in his name.

The main question raised by this exhibit at the Panopticon is why most modern accounts steadfastly refuse to mention that the entire affair resulted from a fantastic coincidence. The tale has been in the public domain for a century and a half; it is hardly top-secret. Perhaps writers of textbooks and encyclopedias feel the controversy, mistake versus prediction, is still open despite all evidence. Perhaps, unconsciously, these authors intend to perpetrate the crowning glory of science, its predictive power. In either case, it is an example of doublethink. It is exactly science's predictive power—or maybe in this situation its retrodictive power—that proved the discovery to be an accident. Accidents happen, and as we have seen and will see throughout our tour, science relies on them. To pretend otherwise is disingenous.

Lastly, a postscript. Neptune may have been done, but not Le Verrier. He eventually turned his attention to Mercury. It had long been known that Mercury's perihelion—the point of closest approach to the Sun—continually advanced at the rate of about 580 arc-seconds per century. After a detailed examination of hundreds of observations, Le Verrier announced in 1859 that when he subtracted the gravitational influence of all the planets on Mercury, 38 arc-seconds per century were left over, due to "some as yet unknown action on which no light has been thrown . . . a grave difficulty worthy of attention by astronomers." An amateur's mistaken observation convinced Le Verrier that a new planet interior to Mercury, Vulcan, was responsible for this tiny anomaly. Vulcan's existence was never confirmed, but Le Verrier continued to believe in it until his death on September 23, 1877, thirty-one years to the day after Neptune's discovery. This 38 seconds of arc, later amended to 43, was Le Verrier's real epitaph, for the inordinately tiny number remained the single glitch in the Newtonian world clock for over sixty years and foreshadowed an entirely new theory of gravity. But let us recount that story, of Einstein's general relativity, in its proper place.

5 / Invisible Light: The Discovery of Radioactivity

It is perhaps, along with Fleming's discovery of penicillin, the most famous accidental discovery in science, the best demonstration that chance favors the prepared mind. The dogged researcher has learned that after exposure to sunlight, a certain mineral containing uranium—a uranium salt—blackens a photographic plate wrapped in dark paper. Was the uranium salt somehow storing the sun's rays? As he is about to repeat the experiment, the weather in Paris turns cloudy, and the researcher puts the unused plate with the uranium crystal atop it into a drawer. Several days later, impatient with the weather, he decides to develop the plate anyway and—to his eternal astonishment—finds it blackened without any prior exposure to sunlight of the mineral. The uranium salt has been emitting invisible and hitherto undiscovered rays, which have the ability to penetrate paper and expose photographic emulsion.

Henri Becquerel, of course, discovering radioactivity in 1896. No, Abel Niepce de Saint-Victor discovering radioactivity in 1857.

One wonders what thoughts were running through Michel Chevreul's mind when in 1842, as head of a commission appointed by the Ministry of the Army, he traveled to Montauban charged with evaluating a proposal on the dyeing of uniforms made by one lieutenant of the dragoons, Claude-Félix-Abel Niepce de Saint-Victor.

One imagines the lot of them meeting in an abandoned factory on the riverbank, or perhaps in an old warehouse, peering into vats of dye, or examining with serious visages the uniforms colored by the new process. We don't know any of this. We don't even know whether Chevreul actually traveled to Montauban. We know only that Michel-Eugène Chevreul, already fifty-six and director of dyeing at the Gobelins national tapestry manufactory, had been appointed by the Ministry of the Army to evaluate Niepce's proposal, whatever it may have been, and that Niepce was then stationed at Montauben. The lieutenant, a nephew of the famous Nicéphore Niepce, who had taken the first permanent photograph, was then about thirty-seven, having been born on July 26, 1805, in Saint-Cyr, near Chalon-sur-Saône. Like many of the time he seemed destined for a military career, and in 1827 entered calvary school; by 1842 he had become a lieutenant in the first regiment of dragoons at Montauban. We don't know what he looked like—if a portrait exists, it has long been buried. It is pleasant to picture him as a dashing officer outfitted in brilliant blue and red, carrying a saber and polishing his dress hat. But it is more difficult to imagine that either he or Chevreul had any inkling of the great and perplexing discovery they were to make, the controversy it would cause, or that the world would utterly forget it.

Regardless of what actually took place in 1842, we know that in 1845 Niepce received a transfer to Paris, where he could work under the direction of Chevreul, who was a professor of chemistry at the Museum of Natural History and a member of the Academy of Sciences. Given that Nicéphore Niepce had taken his first photograph only three decades earlier and Louis Daguerre had announced his daguerrotype in 1839, it is hardly suprising that Chevreul had an interest in photography, even less Niepce de Saint-Victor.* Chevreul encouraged his protégé to work on the development of new photographic processes and emulsions, and even today Niepce is credited as having been first to use albumen in photographs and the first to produce negatives on glass (sometimes one hears steel, maybe both).

Photography is, if nothing else, the interaction of chemical substances with light, and we can be sure Niepce investigated dozens, if not hundreds, of variations over the next decade. By 1856 he was sufficiently astonished by the behavior of certain chemicals to write a memoir, "On a New Action of Light," which was communicated by Chevreul to the French Academy of Sciences on November 16, 1857, and published in the *Comptes Rendus* of the academy. Niepce begins:

> Does a body, after having been struck by light, or having been subjected to exposure, retain in darkness some impression of this light? This is the problem that I have tried to resolve through photography. Phosphorescence and fluorescence of bodies are known: but, as far as I am aware, no one before me has ever done the experiments I am about to describe.

---

* The history of photography, of course, is its own tangled mess. Thomas Wedgewood in England reported to the Royal Institution as early as 1802 that he was able to make a silhouette of a leaf on sensitized paper, but that the image faded. In England, Fox Talbot is regarded as the father of photography because he made his earliest exposures in 1834. Daguerre, however (apparently knowing of Talbot), announced his process to the public first, in 1839, and so in France, Talbot is never mentioned.

What Niepce had been doing was to expose plates covered with a variety of chemicals to light, then cover the plates with photographic paper and put everything in a dark drawer for several days. To his amazement, he found that certain chemicals exposed the photographic paper in absolute darkness. As both the title of his memoir and the opening paragraph make clear, Niepce was convinced he had discovered something new. They also make clear that, initially, he believed that this new phenomenon had something to do with exposure of the substances he was testing to light. But what exactly? Fluorescence, the ability of certain chemicals to glow under exposure to light, had been known for hundreds of years. Phosphorescence, the ability of some substances to emit light after exposure has ceased, had been observed at least since the seventeenth century. Niepce understood he was witnessing neither—no visible light was being emitted. In which case, what?

One reads Niepce's memoirs with a sense of frustration. Looking over his shoulder, you can't help but advise, "Stop exposing these plates to light; it has nothing to do with that!" But hindsight has had the benefit of Lasik surgery. In the thick of things one never has any idea of what is going on, then as now, and every insight is territory claimed by blood. Niepce would never get over the idea that this new phenomenon must be some sort of stored light, but he did make progress. By the second memoir, presented to the academy in 1858, he realizes that if one wants to get a quick and "vigorous" image, then one should impregnate a piece of paper with uranium nitrate—exactly one of the uranium salts Becquerel would use forty years later. He is still first exposing the uranium nitrate to the sun, and even thinks that tartaric acid works as well, which is nonsense. (One can only guess that he contaminated his tartaric acid with uranium nitrate.) Nevertheless, Niepce is able to say, "In conclusion, I have observed that the bodies which conserve the most activity after exposure are those, with the exception of the uranium salts, that are the least fluorescent."

Despite the confusion, director and researcher gradually began to perceive the importance of the discovery. After Niepce's third memoir to the academy, Chevreul presents his own. "The facts set forth in M. Niepce's last memoir are important," he says, "not only for their connection to questions concerning chemical phenomena produced or assisted by the action of light, but also because above all they are new and concern *light itself, its dynamical force.* This is a capital discovery. . . . M. Niepce has established that the exposed cardboard, kept in a tin cylinder in darkness is still active six months after exposure."

And finally, in 1861, Niepce states conclusively, "This persistent activity . . . cannot even be phosphorescence, because it would not continue for such a long time, as has been shown by Edmond Becquerel. It is then more probable that this is some sort of rays invisible to our eyes, as M. Léon Foucault believes, rays that do not cross glass."

That Foucault, one of the most eminent scientists of the time, commented on Niepce's findings indicates that they stimulated some interest. To be sure, Niepce presented no fewer than six memoirs to the academy between 1857 and 1867. In one he notes that "the action of light is perhaps very favorable on certain wines; it can give them the quality of an old wine." In a few he responds to his critics: "Many hypotheses have been given. Certain people have denied the facts, which is the easiest thing to do. But no one has given an explanation of this phenomenon." Edmond Becquerel (1820–1891), Henri's father, who worked at the Museum of Natural History with Chevreul, also commented on the discoveries and gave his own, incorrect, explanation.

One would have thought that with such prolonged replication of the findings, other researchers would have followed up Niepce's claims. Perhaps the reports were too fantastic to be taken seriously, so far ahead of their time that no one had a conceptual framework in which to fit them. For whatever reasons, Niepce's facts were denied, then forgotten. Niepce died in 1870, "leaving for posterity a widow, two children and nothing!" Chevreul followed him nineteen years later, at the extraordinary age of 103. Perhaps his longevity had something to do with the action of light on wine. But he was gone, and that left the stage empty for the appearance of Henri Becquerel.

It is easier to imagine what was going on in Henri Becquerel's mind when in 1895, three years after he had taken his father's chair at the Museum of Natural History, he read William Conrad Röntgen's paper on the discovery of X rays. Röntgen had been experimenting with cathode ray tubes, also known as Crookes tubes or Geissler tubes, depending whether you are English or German, which are no more than glass tubes partially evacuated of air and through which high-voltage electricity can be discharged. Today we call them television tubes; cathode rays are electrons. Since about the time Niepce submitted his second memoir to the academy, scientists had known that cathode rays striking the end of such a tube could cause it to phosphoresce. Röntgen, in another of history's famous accidents, discovered that the cathode rays striking one end of the tube were causing it to emit invisible radiation, which passed through cardboard and air and humans. The forty-three-year-old Becquerel, who, like his father, had an interest in studying solar spectra with photography, reasoned that certain substances struck by light might also emit invisible radiation. He exposed a uranium salt to light; the weather in Paris turned cloudy. The rest is history.

Becquerel, it needs to be said, does not come off well in the remainder of this drama. Shortly after he announced "invisible radiation emitted by uranium salts" to the academy in 1896, several respected scientists pointed out that the same discovery with the same mineral and virtually the same methods had been made forty years earlier, and published in the same journal. One eminent critic, Professor de Heen of the University of Liège and a member of the Belgian Royal Academy, went so far as to publish a pamphlet in 1901 asking, scandalized tone audible between the quotation marks, "Who is the discoverer of the phenomenon called radioactivity?" Despite the clamor Becquerel refused to mention his predecessor for seven years. When he finally did, in 1903, it was only to dismiss Niepce's work. "Uranium is in such small quantities on those papers," he wrote, "that in order to produce an appreciable effect on the [photographic] plates used by the author, it would have been necessary to keep them in place several months. Therefore Niepce could not have observed rays of uranium."

To make matters worse, Becquerel rewrote the history of his own contributions. In reading his first memoirs to the academy, it is quite clear that Becquerel, like Niepce before him, initially thought that he had discovered some sort of *lumière noire,* which could be reflected, polarized, and refracted (shades of Young) just as ordinary light could be. (At the same time, though, it seemed to have electrical properties.) Becquerel's most vocal critic, one of those who pointed out Niepce's priority in the matter, was physician and sociologist Gustave Le Bon. It appears to have been Le Bon's experiments in the years 1896–1899 that actually showed that the new rays did not have the properties of light but were identical to cathode rays, that is, electrons. (Certainly Le Bon is of this opinion.) And that meant Becquerel's first experiments must have been in error. Becquerel came around to this point of view himself by 1899 with no acknowledgment of Le Bon. Indeed, in one of his own popular books, Becquerel charged that "it is sufficient to read M. Gustave Le Bon's work in the *Comptes Rendus* to convince oneself that the author . . . had no idea of the phenomenon of radioactivity." Le Bon was not amused. Supporters of Becquerel continued to claim that he had shown that uranium rays were neither reflected nor refracted.

Apart from all this, the nagging question remains of whether Henri Becquerel had known of Niepce's work and had conveniently forgotten it (Kekulé seems to have excelled at that sort of thing; see chapter 18). There can be no question that Edmond, who worked with Chevreul at the museum, knew of the discovery. One of Edmond's memoirs on phosphorescence in fact appears directly after Niepce's first report in the *Comptes Rendus*. Le Bon, no friend of Becquerel's, claims that his opponent's initial misunderstandings about the discovery were due to the fact that he was "always under the influence of the ideas of Niepce de Saint-Victor." On the other hand, Henri's son, Jean, no neutral observer either, always claimed that the path of discovery went from Edmond to Henri and credit should, above all, go to the museum. All in all, there seems to be no hard evidence that Edmond told his son about Niepce's work, and such accusations should be consigned to the armchair.

Niepce's discovery, of course, was before its time. It is convenient for that reason to nod with sympathetic understanding that it would inevitably find the Oblivion File. By 1896, we understand: Crookes tubes were commonplace, X rays had been discovered, and the academicians were more receptive to the idea that Becquerel had uncovered something new. Of course it was new; this was not stored light, but radiation emitted from the disintegration of the atomic nucleus itself. It portended not only an entire new branch of physics but a new danger to mankind.

Yet there was more to this sad tale than a discovery made forty years before it should have been. Niepce's work attracted enough interest at the time that it might have been pursued; by then the action of cathode rays on glass had been noticed. History could have been otherwise. In any case, the clamor raised by Le Bon at the turn of the century should have ensured that Niepce's name would be resurrected. It was not. Another factor, an inescapable one, clearly contributed to Niepce de Saint-Victor's perpetual interment. One of the largest exhibits at the Panopticon, and certainly the most colorful, is devoted to this phenomenon, the Nobel prize. Once Becquerel had been awarded the prize in 1903 with his student Marie Curie and her husband, Pierre, that was the end of the story. Nobel prizes are given for discoveries, not always for understanding, and they have had the effect of clamping history shut. As strange as it may seem, in our own time no story more closely parallels the discovery of radioactivity than the discovery of the cosmic microwave background radiation. The accidental detection of the background radiation by Arno Penzias and Robert Wilson at Bell Labs in 1964 is considered one of the twentieth century's supremely important cosmological observations, for it convinced cosmologists that the big bang theory was correct. The microwave background is nothing less than the heat left over from the big bang.

Yet it had been predicted as early as 1950 by Ralph Alpher and Robert Hermann, who didn't take their own prediction seriously enough to follow up. At least three research groups around the world independently repredicted it in the early 1960s; two Princeton cosmologists were completing the apparatus to search for it in 1964, and when Penzias and Wilson stumbled across the signal, the Princeton group explained to them what they had found. This does not count all the spurious, near, and ignored detections over the previous two decades. Physicists know the story, but in textbooks it is reduced to a single line. Who deserved the prize?

Many years ago I remarked to Tullio Regge, an Italian physicist, that the Nobel prize should be abolished. Regge countered that that would only make the prizes in circulation more valuable. Rather, the Nobel prize should be inflated. The committee should hand out more and more every year until the prize was worthless. During that memorable drive with Feynman to the lake the subject, inevitably, turned to Nobel prizes, and I mentioned Regge's suggestion. Feynman, in his usual growl, rejoined, "Regge just said that because he's an Italian. He's always thinking about inflation. Tell me, kid, what do you think about *my* Nobel prize?"

That's another story, but one thing Abel Niepce de Saint-Victor's tale teaches us is that the history of science should be divorced from the history of the Nobel prize. It may be that justice will eventually have its say: Over the last few years Abel Niepce de Saint-Victor seems to have become recognized by a few people in France as the discoverer of radioactivity, but the news has yet to cross the Atlantic. We can only hope it someday makes the journey.

6 / Light, Ether, Corpuscles, and Charge: The Electron

At a recent symposium on the play *Copenhagen,* a Harvard Nobel laureate informed the audience that the electron was discovered in 1897 by J. J. Thomson. It must be true; every textbook says exactly the same thing, and in 1997 the American Association for the Advancement of Science held a centennial celebration to prove it. Somewhere I must have taken a wrong turn. But I am not alone. The arrival of the first particle, the beginning of all microphysics—no event was of greater import for the coming revolution than the discovery of the electron. No event in the last few years has come under greater historical scrutiny. And in no case is the usual telling more incomplete.

In the usual telling, the tale is fairly just so. The crucial measurement that convinced the scientific community of the existence of electrons was the determination of the fabled charge-to-mass ratio of cathode rays, the same cathode rays that so perplexed investigators in the previous exhibit. If one believed that cathode rays were composed of invisible, electrically charged particles, then it seems eminently reasonable that one should be able to measure the ratio of their charge to their mass; indeed, it turns out to be much easier to measure this ratio than either the charge or the mass individually. When in 1897 Thomson finally established that the charge-to-mass-ratio of cathode rays was about two thousand times larger than for hydrogen atoms, he was faced with a choice: either the new particle had a charge two thousand times larger than hydrogen or a mass two thousand times smaller, or perhaps some combination of the two. Thomson decided that cathode rays were composed of particles with a mass two thousand times smaller than that of a hydrogen atom. Thus was born the electron.

The true story is better. In fact, there is no single "story." More than any other exhibit in the Panopticon of Present and Past Concepts, the advent of the electron raises this issue: What does it mean to discover? Electrons cannot be seen with the naked eye, and all their effects must be observed indirectly. Detectives attempting to nail such a perp must rely entirely on circumstantial evidence. When is the case solved? When the investigator first has a hunch that eventually pans out? When you, the reader, guess the ending? When the master detective gives his wrap-up speech at the finale, explaining why it all fits together? When you close the book, convinced?

In the case of the electron the evidence for its existence accumulated so gradually and from so many different quarters that it is difficult to assign a date to the event of its discovery—if "event" it was. Some historians balk at the notion of calling it a discovery at all. In 1897 the various investigations reached a climax when at least three scientists simultaneously announced the charge-to-mass ratio of the new particle. Thereafter the idea of the electron gradually became accepted, as the last of its opponents died off.

But let us start from the beginning, wherever that is. Remember, in the early nineteenth century the notion of atoms is not yet firmly fixed in the minds of natural philosophers. John Dalton in 1803 has resurrected the Greek idea of atoms; around 1811 Avogadro and Ampère propose that equal volumes of gases contained equal number of molecules, but Avogadro's hypothesis remains unaccepted by chemists for a full half century, which gives a good indication of how difficult the concepts are to swallow.

For our purposes we might date the beginning of the electron to Faraday's 1833 experiments on electrolysis, experiments that Joseph Henry does not appear to have duplicated. Faraday observed that when he passed an electric current through a solution of, say, hydrochloric acid, the acid would separate into hydrogen and chlorine, and the hydrogen would migrate toward the negative electrode and the chlorine toward the positive electrode. This process, termed electrolysis, "to loosen with electricity," is the basis for all chemical batteries. What Faraday realized was that the amount of a substance deposited on an electrode was directly proportional to the total amount of electricity he passed through the solution. One amp of current for one hour would deposit so many grams of chlorine; ten amps for an hour—or one amp for ten hours—would deposit ten times that amount. Furthermore, for a fixed quantity of electricity, the mass of a substance liberated by electrolysis was directly proportional to what we call its atomic weight (the mass of an element compared to hydrogen).

If one imagines that the electric current in Faraday's solution is composed of discrete charges moving between electrodes, then his findings are reasonably easy to interpret. A fixed amount of electricity contains a fixed number of charges. Only this number of chlorine ions can pass between electrodes, and the same number of hydrogen ions.* But since chlorine weighs thirty-five times as much as hydrogen, the mass of chlorine deposited on the anode should be thirty-five times that of hydrogen.†

None of this was so clear at the time. To natural philosophers of the day electricity was a "galvanic fluid." Nevertheless, Faraday glimpsed the truth. "Although we know nothing of what an atom is," he wrote on the last day of 1833, "yet we cannot resist forming some idea of a small particle, which represents it to the mind; and though we are in equal, if not in greater ignorance of electricity, so as to be unable to say whether it is a particular matter or matters, or mere motion of ordinary matter, or some third power or agent. . . . there is an immensity of facts which justify us in believing that the atoms of matter are in some way endowed or associated with electrical powers, to which they owe their most striking properties." Then, as if the truth were too incomprehensible, he stepped away from it: "I must confess I am jealous of the term *atom;* for though it is very easy to talk of atoms, it is very difficult to form a clear idea of their nature, especially when compound bodies are under consideration."

Faraday's reluctance to speculate was legendary, but other natural philosophers displayed less reticence, and by the 1840s brave souls were postulating that the electric fluid actually consisted of a stream of moving charges. In 1874 the prophet George Johnstone Stoney announced on the basis of electrolysis experiments a value for the elementary unit of charge, which was only about a factor of fifteen off from today's value. Sometime after 1881 he began using the term *electron* for this fundamental unit, a term that rapidly grew in popularity. Nevertheless, one should not confuse Stoney's unit of charge with the modern notion of the particle. But he was remarkably close to the mark: He believed that the motion of electrons in the ether produced light. Which is true.

More or less. The ether, the ether. Let's talk about the ether later. Further evidence for the existence of discrete charged particles came at the end of the 1880s from an entirely different direction—the photoelectric effect. The photoelectric effect refers to the ability of zinc and other metals to produce an electric current when struck by light, and the phenomenon is

---

* An ion is an atom that has had one or more electrons removed or added. As a result, the atom becomes charged and can be attracted or repelled by a charged electrode.
† We eschew complications due to valence.

the basis of many devices we take for granted today, including electric eyes and solar cells. As is widely known, Heinrich Hertz discovered the photoelectric effect in 1887 as a by-product of his discovery of electromagnetic waves.

In a series of experiments remarkably similar to those performed by Joseph Henry fifty years earlier in Princeton (chapter 3), Hertz found that sparks from a high-voltage coil in one circuit could cause sparks to jump between two wire tips in a totally disconnected circuit across the laboratory. Hertz is justly renowned for the discovery and went considerably farther than Henry in confirming that these new disturbances traveled at finite velocity (which as far as he could tell was about the speed of light), were reflected like light waves, and underwent diffraction and interference like light waves. In fact, they were light waves—more precisely, radio waves.

It is rather extraordinary, in light of the vicious debate over the invention of the radio (chapter 14), that no one has mentioned how seven years earlier Hertz's discovery was anticipated by a music professor, David Edward Hughes (1830–1900), who showed that signals sent from a spark gap transmitter could be received up to five hundred yards away by his primitive microphones connected to a telephone earpiece. In 1879 and 1880 he demonstrated these experiments before the president and several fellows of the Royal Society, correctly arguing that the signals were being transmitted by electromagnetic waves. Thus Hughes might be said not only to have discovered radio waves but to have built the first radio receiver. Unfortunately, the judges decided that the effect could be explained by known physics.* Disheartened, Hughes refused to publish an account until 1899, by which time he was on the verge of perishing. Even more amazing in its neglect is that this well-documented story can be found in Edmund Whittaker's *History of the Theories of Aether and Electricity,* the bible on nineteenth-century electromagnetic theories.[†]

Ah, the photoelectric effect. During the course of his researches, Hertz darkened the room in order to observe the dim secondary spark and to his surprise found that eliminating stray light caused the spark to become weaker. Following his nose, he interposed sheets of various materials between the primary and secondary sparks. Indeed, this also weakened the secondary spark. Evidently light from the primary spark enhanced the secondary. He moved the secondary farther away from the primary. The secon-

---

* Their argument was the same as some critics use to dismiss Henry's earlier results—a magnetic field *induced* a current in the receiving circuit, with no waves involved. The point is an extremely subtle one.
[†] The author has posted Hughes' account on his Web site (wwwrel.ph.utexas.edu/~tonyr).

dary spark disappeared. He shone a nearby arc lamp on the secondary spark gap; the sparks reappeared. Soon enough Hertz determined that ultraviolet light produced the most vigorous sparks.

Hertz did not pursue the matter further, but his assistant Wilhelm Hallwachs (1859–1922) did and showed that the light hitting the metal was causing an electric charge to be given off. Internet sources and even some textbooks credit Hallwachs with the suggestion that the electrical current was due to the emission of individual particles, and by implication credit him with the world's first suggestion of the electron as we know it. This is a slight case of retroactive overinterpretation. What Hallwachs did say was, "One is entitled to assume that when polished electrically negative metal plates are hit by light the surfaces of these plates are changed in such a way that negative electrical particles are given off from them." He then offers a few vague remarks and ends by admitting that "a final conclusion cannot be made."

Hallwachs, then, seems to have been on the right track, but at the very least the charges he was thinking of were ions or gas molecules, not electrons. Still, although the photoelectric effect is commonly credited to Hertz, the electric half *was* due to Hallwachs, and in some older sources it is termed the "Hallwachs effect."* The phenomenon proved to be one of the great mysteries of fin de siècle physics. All attempts to understand it in terms of contemporary theories of light failed miserably, and the matter was only cleared up in 1905 by Einstein himself.†

Physics does not cease at the tenth meridian. Italians point out that the eminent scientist Augusto Righi of Bologna (whose name will arise again) simultaneously carried out experiments similar to those of Hallwachs, and one should not forget that Aleksandr Stoletov in Moscow independently stumbled onto the effect at about the same time. Stoletov (the name means "man of one hundred years"), as it turns out, was a leading figure at Moscow State University, did much to organize Russian science, and was the first to actually construct what we might call a photocell. Unfortunately, according to some, he ran afoul of the powerful Count Golytsin, an amateur scientist, was dismissed from the university, and committed suicide at the age of fifty-nine, forty-one years short of his allotment. On the other hand, in standard Soviet versions of the story, he dies of natural causes.

---

* Britannica actually mentions an 1839 sighting of the photoelectric effect but fails to provide citations.
† For a more complete discussion of the photoelectric effect, see *Doubt and Certainty* by the author and George Sudarshan (Reading, Mass.: Perseus, 1998).

*   *   *

We must now return (briefly) to the Panopticon's exhibit on the ether, which we visited in chapter 2. Since Newton, natural philosophers had usually conceived the ether as the all-pervasive medium that allowed forces, such as gravity, to propagate through space. After James Clerk Maxwell published his theory of electromagnetism in the 1860s, it became the medium that allowed the propagation of electromagnetic waves. "Take electricity out of the world and light vanishes," Hertz declared. "Take the luminiferous ether out of the world, and electric and magnetic forces can no longer travel through space."

Maxwell's theory, as successful as it was, had a few defects. One of them is that it took no account of how material bodies interacted with the all-pervasive ether. For example, when electric charges move, they constitute a current, and currents generate magnetic fields, which are transmitted through space. How? Theories abounded, myriad theories as arcane as string theory seems today. Vortexes were big. The ether consisted of vortexes; atoms and electrons also became vortexes in the ether. Lord Kelvin himself introduced a model called the vortex sponge.

Why even talk about such theories a century after they were all thrown into the dustbin of history? In an attempt to tread carefully. When Sir Joseph Larmor announced his theory of electrons in 1894, they had the same mass as hydrogen atoms but were structures in the ether with "vacuous cores round which the radial twist is distributed." Whatever this is, it is assuredly not our electron, and illustrates just how foreign the conceptual framework of those men was to our own. It seems nearly a miracle that any of those elaborate constructions gave sensible answers.

At least one did. In 1892 Hendrik Antoon Lorentz (1853–1928), the greatest physicist of his time, introduced his theory of electrons, which attempted to describe all electromagnetic phenomena as a consequence of moving electrons embedded in the ether. Lorentz's electrons were not our electrons. They were merely charges—at least until 1895, when he identified them with the "charges of electrolysis," or hydrogen ions. Although the theory would ultimately turn out to be incorrect, in it Lorentz introduced his famous law that correctly describes how charges move in a magnetic field.

The plot thickens. Pieter Zeeman (1856–1943), Lorentz's assistant at the University of Leiden, had decided to investigate the effect of magnetic fields on light, an area Faraday had pioneered decades earlier, and to see "whether the light of a flame if submitted to the action of magnetism would undergo any change." Zeeman's first experiment was so simple that it seems incredible it led to a great discovery: He held a piece of asbestos impregnated with common table salt above the flame of a Bunsen burner, which had itself been placed between the poles of an electromagnet. Most of us who have sprinkled salt into a flame know that it glows bright yellow. The color is actually due to two closely spaced spectral lines, which are made clearly visible by an ordinary prism. When Zeeman examined the burning salt, sure enough, the lines were there, narrow and sharp. When he turned on the magnet, they spread apart. Zeeman could hardly believe it and performed several more sophisticated experiments to convince himself the effect was real. It was.

His mentor, Lorentz, provided an explanation. His electron theory predicted the spreading of the spectral lines, as well as several other properties, all of which Zeeman confirmed by experiment. It also allowed one to calculate e/m, the charge-to-mass ratio of the ions responsible for the Zeeman effect. In the October and November 1896 issues of the *Verslagen* of the Amsterdam Academy, Zeeman announced his results:

> The experiments . . . may be regarded as a proof that the light-vibrations are caused by the motion of ions, as introduced by Prof. Lorentz in his theory of electricity. From the measured widening . . . the ratio e/m may now be deduced. It thus appears that e/m is of the order of magnitude 10^7 electromagnetic C.G.S. units. Of course this result from theory is only to be considered a first approximation.

In plain language he is saying that e/m for Lorentz's ions is about a thousand times larger than for hydrogen. Yet this figure—which to us immediately suggests that the particles producing the Zeeman effect must be a thousand times lighter than hydrogen—goes by without comment. In his next paper, presented before the academy in June 1897, Zeeman does remark, "It is very probable that these 'ions' differ from the electrolytical" and "The high value of e/m which I have found makes it extremely improbable that we have to deal with the same mass in the two cases." In a third paper, delivered a month later, Zeeman gives a refined value for e/m, which (assuming a misprint, as the printed value apparently makes no sense) is almost exactly today's value. But again he makes no comment on the number. All these papers appeared in English translation in the March, July, and September 1897 issues of one of the preeminent scientific journals of the day, the *Philosophical Magazine*. The last was a month before Thomson's famous paper appeared in the same journal.

Meanwhile, cathode rays. Since their discovery in 1857, cathode rays had been extensively studied. Scientists knew that when a high-voltage supply was attached to electrodes embedded in evacuated glass tubes, cathode rays were given off by the negative electrode; they could be deflected by magnetic fields, and they could even push little pinwheels. They did not know what constituted cathode rays. As far as many physicists were concerned, especially the Continentals, they were disturbances in the ether.

Hertz himself was detoured in the 1880s by some false results that led him to believe that cathode rays were uncharged etherial ripples. But in 1891 he discovered that cathode rays penetrated thin gold foil. He planned to use this fact to test again whether the rays were charged, but in order to do so he needed to find a way to allow them to exit the tube. Unfortunately, he was by then extremely ill, and it was left to his assistant Philipp Lenard to complete the experiments after Hertz's death. Lenard covered one end of the tube with a thin foil, which allowed the cathode rays to escape. He was then able to learn that the rays were deflected by electric fields and to ascertain that they were not composed of molecules. Neither were the French sleeping. In 1895 Jean Perrin actually managed to collect the charge carried by cathode rays and establish that it was negative. Obviously the rays were composed of ions, which were pieces of residual gas molecules in his vacuum tube that had been broken up by the intense electric field.

To the provinces. Emil Weichert, a twenty-six-year-old *Privatdozent* (an unpaid adjunct faculty member) at the University of Königsberg, was also experimenting with cathode rays. In a lecture on January 7, 1897, he reported his results and concluded, "We are not dealing here with atoms as we know them in chemistry. I can say this because the mass of the particles set in motion can be shown to be 2000 to 4000 times smaller than the mass of the hydrogen atom—the lightest of all chemical atoms." Weichert referred to these new entities as "electric atoms," and it seems to be the first time anyone ever publicly referred to a subatomic particle. His estimate for the mass was not bad either, and relied on the assumption that the charge was one Stoney electron.

In Berlin at the same time, twenty-five-year-old Walter Kaufmann was pursuing cathode rays. His experiments were carried out with great care and in a paper finished in April 1897 and published in the June issue of *Annalen der Physik,* the leading German journal, he gave an estimate of e/m that was exactly Zeeman's first number. In a second paper, written with E. Aschinkass and published in November, he arrived at a value for e/m that is in virtual perfect agreement with today's value. The results evidently disturbed Kaufmann more than they had Zeeman. Although he observes, "The fact that cathode rays are deflected by magnetic fields has often been taken as proof that they are negatively charged particles," once he has his results in hand he bends over backward to explain them away and ends by writing, "I believe I am justified in concluding that the hypothesis of cathode rays as emitted particles is by itself inadequate for a satisfactory explanation of the regularities I have observed."

And in December, Philipp Lenard submitted a similar value to the same journal. But Lenard still seemed wedded to the possibility that he was observing "some special parts of the ether which have thus far escaped notice." Anyway, by this time it was too late: Lenard was forced to acknowledge that J. J. Thomson had beat him to the punch.*

The details of J.J.'s experiment need not concern us; the setup can be found in any textbook and is a classic. With it Thomson measured the deflection of cathode rays in a magnetic field and was able to deduce e/m. What strikes one in reading his famous 1897 paper, which appeared in the October *Philosophical Magazine,* is how readily, in distinction to the Continentals, he seems willing to accept the idea of a particle. "The electrified particle theory," he writes with prescient distaste for the competition, "has . . . a great advantage over the aetherial theory, since it is definite and its consequences can be predicted; with the aetherial theory it is impossible to predict what will happen under any given circumstances, as on this theory we are dealing with hitherto unobserved phenomena in the aether, of whose laws we are ignorant." With even more dripping condescension, he continues: "Now the supporters of the aetherial theory do not deny that electrified particles are shot off from the cathode; they deny, however, that these particles have any more to do with the cathode rays than a rifle-ball has with the flash when a rifle is fired."

Thomson was a relative newcomer to cathode ray research, and his strong convictions are a little surprising. His main line of research had actually concerned electrical discharges in gases, and although he advocated the "corpuscular" nature of cathode rays, even in 1895 these corpuscles arose from elaborate vortex-gyroscopic models that connected atoms with the ether. What caused him to get into cathode rays is unknown; almost certainly Perrin's discovery that the rays were negatively charged

---

* Lenard is difficult. Although he was later to win the Nobel prize for his work on cathode rays, he was such a disgusting character that many physicists have gone out of their way to completely discredit him. He was a virulent anti-Semite, openly attacked Einstein, and in old age volunteered to lead Hitler's campaign against Jewish science. Arnold Sommerfeld credits Hertz with the conception of the foil window (the "Lenard window") on the discharge tube, but it does seem Lenard carried out the experiments, and I don't know any reason to deny that in 1897 he found a value of e/m simultaneously with Thomson and Kaufmann.

played a decisive role. So by 1897 Thomson was well primed for some sort of particle, but his conception of these particles—which he took to be the unique building blocks of matter—corresponded more closely to our concept of atoms than electrons.

If one insists, as many physicists do, that the measurement of e/m constituted the discovery of the electron, then it is difficult to call Thomson the discoverer. Clearly Zeeman did that first, and Thomson's values are not even as good as Kaufmann's, who did it simultaneously. Indeed, Thomson closed his April 30, 1897, lecture, in which he first announced his findings, by noticing that his value for e/m agrees with Zeeman's. On the other hand, Zeeman's results were based on a particular theory, and it perhaps wasn't obvious to everyone that he and Thomson were talking about the same thing. In any case, numbers without interpretation are meaningless. Zeeman clearly balked at interpreting the data before his eyes; neither Kaufmann nor Lenard go the full distance. Thomson, on the other hand, is willing to take the plunge. To him the high value of e/m he has obtained implies particles, no ifs, ands, or buts.

Still, one can counter that Weichert was the first to propose a mass for a new particle. Furthermore, Kaufmann, in his interesting 1901 lecture "The Development of the Electron Idea," gives the impression that Thomson was just one of many who measured e/m and that once everyone realized the results were the same as Zeeman's, "the hypothesis probably first put forward by Weichert may be unhesitatingly adopted: that we have in both cases the same particles, viz., the electrons." In other words, by 1897 it was obvious. (At least it was obvious in 1901 that it was obvious in 1897.)

One must also concede that J.J. did *not* in 1897 conclude that the mass of the electron was about a thousandth that of the hydrogen atom. Rather, "the smallness of the value of m/e is, I think, due to the largeness of e and the smallness of m." It would take another two years of innovative experiments to convince Thomson that the mass of the new particle was indeed tiny. However, once he identified his "corpuscles," he went on to propose a model of the atom, which none of his rivals did. This atom, though, to complicate things further, was not our atom (having no nucleus but only corpuscles in a positively charged background), and in fact it was *not* Thomson who suggested that his corpuscles were what we would call electrons. That proposal seems to have come from Joseph Larmor, who after Zeeman's experiment had arrived at a reasonably modern conception of the particle.

Perhaps armchair partisans regard these final steps as after the fact. Certainly contemporaries did; within a few months of Thomson's 1897 paper, except for a few diehards, opposition to the idea of electrical particles crumbled.

Despite the tortured path to the electron and the photo finish, Thomson is still remembered as the sole discoverer of the first particle. If you have any doubts, check the issue of *Physics Today* devoted to the centennial of the electron; of all the contenders who might lay claim, exactly one is mentioned.

Why has history been reduced so in the community that should be most interested in it? There can be little doubt that Thomson's papers are the most decisive. Furthermore, he was the head of a large laboratory, and a coterie of students could pass on the legend. It could not have hurt that in 1907 Sir Oliver Lodge published an influential book *Electrons,* which, although acknowledging European contributions, traced the history almost entirely from the English perspective. Both Thomson and Lodge also rearranged a few dates in their favor, which led to a polite rebuttal from Zeeman but in general to a permanent reordering of chronology. And just as in the case of radioactivity, Thomson's 1906 Nobel prize put an unbreakable seal on history, though the Curies actually nominated both Kaufmann and Thomson for the award and Thomson's citation nowhere mentions the electron. The prize was awarded for studies in gaseous discharges. As a final irony, Thomson himself refused to adopt the term *electron* until 1911 or 1912, long after everyone else did. Perhaps there was still some confusion in his mind as to exactly who had discovered what.

That confusion should remain.

7 / Einstein's Miraculous Year (and a Few Others)

For a physicist to write about Einstein is a daunting task. He is the titan who stands at the gates of twentieth-century physics, who changed forever our perceptions of the workings of the universe. He, perhaps alone among physicists, has penetrated the scientific consciousness no less than the public consciousness. And he is in all probability unique in that, despite generations of detractors, he seems to have done virtually everything credited to him. What can one add to the millions of words already written about this giant?

The task is made more difficult for a cosmologist, including this one, one of those physicists who grew up with Einstein's picture on their bedroom wall and who have chosen to follow the path blazed by him—to study relativity, gravitation, and to make what they can of the workings of space and time. To us, Einstein is as close to a hero as we are likely to acknowledge, maybe a demigod, even a god, a benevolent one who could do no wrong. One who, to add to his stature, forged his theories from pure thought, without the crutch of experimental evidence or the shoulders of giants to stand on.

That, at least, was the old view of Einstein. Over the last two decades, primarily due to the ongoing publication by Princeton University Press of Einstein's complete papers and correspondence, the portrait is gradually being redrawn. The new picture of Einstein as it takes shape remains that of a towering genius, but one who was well aware of the science going on around him at the time, one who sometimes grudgingly—if at all—acknowledged his predecessors, and one whose personal relations with women were far from saintly.

We cannot fill in all the gory details, so instead we leave them to the army of journalists who have mined the mother lode for their exposés. What we can do, with luck, is to clear up a few of the scientific and historical misconceptions that still hang in the air after the rain of one hundred million words.

Let us pick up the story in early 1902. Einstein, about to turn twenty-three, has graduated from the Zurich Polytechnic and is in desperate need of a job, not least because he has gotten his fiancée, Mileva Marić, pregnant. He rashly moves to Bern, Switzerland, from Schaffhausen, in the hope of getting a position at the Patent Office. While waiting he learns that Mileva has given birth to an illegitimate daughter, Liserl, whom he never sees. What becomes of Liserl is unknown. During the spring of that year he supports himself by giving private lessons in physics. One of his students, Maurice Solovine, and another, physicist Conrad Habicht, become friends for life, and the three of them form the immortal Akademie Olympia to discuss fundamental issues in physics. Einstein occasionally plays the violin.

Fast-forward to 1905. Einstein got the job and has been a "technical expert, third class" at the Patent Office for three years. He and Mileva, married at the beginning of 1903, have been living as man and wife since, but the union is a melancholy one. Mileva, once intending to become a physicist, twice failed her graduation exams at the Zurich Polytechnic, and the strain of having to give up her daughter is undoubtedly great. Evidently resigning herself to a domestic role, she does not participate in the discussions of the Akademie. She gives birth to a son, Hans Albert, in 1904. Eventually the marriage ends in divorce.

But 1905 is not remembered for any marital friction. It is now, and shall always be, enshrined as the *annus mirabilis,* in which Einstein, infant son on his knee, wrote five papers that changed the face and foundations of physics. To nonscientists it still comes as a surprise that only two of these papers concerned the theory of relativity. The first of the five, and the only one that Einstein ever spoke of as "revolutionary," explained the photoelectric effect.

We have discussed the photoelectric effect in detail in the Panopticon's exhibit on the electron (previous chapter). Here we need only repeat that the effect presented some very strange properties that were totally inexplicable by the contemporary understanding of light.* Einstein accounted for every one of them by a simple hypothesis: Light did not always behave as a wave, as Young had argued, but sometimes as a Newtonian corpuscle (dubbed a "photon" in 1926). Einstein's introduction of the photon (more precisely an "energy packet") was the second use made of the quantum theory, invented by Max Planck in 1900. Although the section of Einstein's paper devoted to the photoelectric effect is a small one, it was for these few pages—not, as is still sometimes stated, relativity—that Einstein won

* See note to chapter 6, p. 57.

the 1921 Nobel prize. To the best of my knowledge no one has ever challenged Einstein's priority on the photoelectric effect, and so we say no
more about it, except that with his explanation Einstein united the wave
and particle pictures of light, which had been at odds for more than two
centuries.

Six weeks after submitting the article on the photoelectric effect, Einstein
completed his second miraculous paper, his doctoral thesis. In this work
he tackled what might today seem to be the mundane problem of the
existence of atoms. This was not a trivial issue in 1905; at that late date—
well after the discovery of electrons and radioactivity—a few eminent diehards, including Max Planck himself, still refused to concede that atoms
existed at all. Here, then, the supremely ironic confirmation of Planck's
own dictum that no theory is fully accepted until the last of its opponents
dies off.

In his doctoral thesis Einstein estimated the size of an average molecule and, using his determination, achieved the best value ever calculated
for Avogadro's number, the number of molecules in one mole of gas.
Believe it or not, it is this paper, not any of those on relativity, that remains Einstein's most frequently cited work. (This is not because it is his
greatest, but because no one reads the original papers on relativity anymore, as will become evident.) What is worth pointing out is that Einstein's method for determining molecular dimensions was rather similar to
the one employed forty years earlier by Joseph Loschmidt, who also wrote
down the structure for benzene before Kekulé dreamed it (chapter 18).
There is no reason to believe that Einstein knew of Loschmidt's work in
1905; he certainly did later. But it must be conceded that Einstein was
consistently vague in his early papers about what he knew and what he
didn't, and afterward made frequently contradictory statements about
such matters. This was particularly true about the genesis of relativity, and
his behavior has led detractors to accuse him of outright dishonesty. One
thing is clear: Einstein's love letters to Mileva at the turn of the century
are unusual for an infatuated young man. Tender endearments compete
and more often than not yield to references to Boltzmann, Mach, Ost-
wald, Maxwell, Hertz, Planck—the great physicists of the age. Einstein
was doing his best to keep up with the literature.

Eleven days after completing his doctoral thesis, Einstein submitted another
article on atoms for publication. This, his celebrated paper on Brownian
motion, is generally regarded as the final "proof" of the existence of atoms,

and it remains Einstein's second most frequently cited work. Investigators had long known that microscopic particles—for example, pollen or sperm—suspended in a liquid jiggled about randomly. After 1828, when botantist Robert Brown published his detailed observations of the phenomenon, proving that not only did "living" particles undergo such motion but inert particles such as smoke and wood did as well, the chaotic jangle of microscopic particles was christened Brownian motion in his honor. Much effort throughout the nineteenth century went into explaining it, and by the 1860s at least three natural philosophers independently suggested that Brownian motion was caused by the collision of the suspended particles with invisible molecules.* This would turn out to be the correct explanation, but at the time it was speculation, which met fierce objections. A legitimate calculation of Brownian motion, not dissimilar in spirit to the one Einstein would perform in 1905, was attempted in 1900 by one Felix Exner. Unfortunately for Exner, he got the wrong answer.

As the story goes, Einstein wrote his paper "blissfully unaware of the detailed history of Brownian motion," a claim that he himself also made to Conrad Habicht. According to this version, he not only predicted the phenomenon but explained it as well. Maybe. To be sure, Einstein omits any mention of Brownian motion in the title of his paper, remarking only in the first paragraph, "It is possible that the motions to be discussed here are identical with the so-called Brownian molecular motion; however, the data available to me on the latter are so imprecise that I could not form a judgement on the question."

On the other hand: From Maurice Solovine we do know that a few years earlier the Olympians had pounced on and devoured the great mathematician Henri Poincaré's enormously influential popularization *Science and Hypothesis.* It is the only book Solovine singles out for comment: "This book profoundly impressed us and kept us breathless for weeks on end." One wonders, then, whether Einstein could have missed the brief passage toward the end in which Poincaré discusses Brownian motion by name and mentions some of the issues it raises. Had Einstein conveniently forgotten what he read? As I have pointed out in regard to Henry and Faraday, it is true, especially for first-rate scientists, that the mere existence of a result is the single most important fact to know. Once the result shines before you, the path often reveals itself. It is quite possible that, having heard of Brownian motion, Einstein needed no further input to explain it. And as I have also said, Einstein was not always forthcoming about what he knew and when he knew it.

---

* Giovanni Cantoni from Pavia and two Belgian Jesuits, Joseph Delsaulx and Ignace Carbonnelle.

In any case, the theory of Brownian motion was independently developed in 1900 by a Frenchman, Louis Bachelier. Bachelier was not actually concerned with the motion of microscopic particles suspended in a liquid. He was concerned with prices on the French stock market. Prices on the Bourse, like particles in a liquid, are subject to a vast array of random forces, so many that the prices' behavior can only be studied probabilistically. This is exactly what Bachelier did in his remarkable doctoral thesis, "The Theory of Speculation." Yet although his paper is couched in terms of futures and stock options and "call-o-more's" (whatever those are), the mathematics is identical to that of Brownian motion, and Bachelier's equation explaining the drift of prices with time is the same as the one Einstein later derived for the position of particles. In his paper Bachelier anticipated the Black-Scholes approach to options trading, and for his prescient work he has in recent years been crowned the "father of economic modeling." At the time, though, Bachelier seems to have been ignored, and he passed into obscurity. Could Einstein have known of his predecessor's work and merely transplanted the mathematics to particles? I am aware of no evidence that this is the case.

It is time to speak of relativity. "Special relativity" is probably the greatest misnomer in the history of science, and before we dip into the tangled genesis of the theory synonymous with Einstein's name, let's stop by the infinite exhibit in the Panopticon that considers the ramifications of the adage "Everything is relative." Einstein never called his creation the "theory of relativity." In the fourth of his miraculous 1905 papers, finished a brief eight weeks after his work on Brownian motion, he referred only to the "principle of relativity," which means there is no absolute standard of rest. All motion is relative. As a consequence, the fundamental laws of nature themselves *must be the same* whether you are standing firmly on the ground or on a train moving at 200 kilometers per hour. To our ears this sounds innocent enough, but all of Newtonian physics is built on the assumption that there *does* exist an absolute standard of rest—the ether—from which you can measure motion. If the principle of relativity is correct, the Newtonian tower of physics must collapse. To be sure, with his completed theory Einstein demonstrated that moving clocks appear to run slow and speeding trains appear to contract relative to you, an onlooker, and in doing so he abolished ancient notions of absolute space and time forever.

However, Einstein's theory also demonstrated that as we change perspective from a moving system to a stationary one many quantities remain unaltered. The whole theory stands not only on the principle of relativity but on the more radical postulate that both you on the ground and your twin hurtling past on an asteroid measure the velocity of light to be *exactly the same:* a little under 300,000 kilometers per second. Nothing relative there. It was evidently Max Planck in 1906 who first used the term "relativity theory," and others followed his lead, but in 1910 the mathematician Felix Klein suggested that the theory be called the "theory of absolutes" (or "invariants"), which would have been at least as appropriate. "Relativity" gradually won out. Einstein surrendered only in 1911, still distancing himself from the term by putting it in quotation marks. One suspects that certain ideas about social relativism would never have achieved such a prominent place in the Panopticon exhibit had the world followed Klein's suggestion. To be sure, in the 1920s, when nonsensical controversies broke out about relativity and relativism, scientists resurrected Klein's proposal, suggesting that the theory be renamed the "theory of absolutes," but by then it was too late.

One should also remark that, contrary to universal belief, Einstein did not invent the idea of the space-time continuum, or simply spacetime, as we now refer to it. Nowhere in his early papers is there any mention of time as the fourth dimension. The joining of space and time into a four-dimensional world was the contribution of the mathematician Hermann Minkowski. Einstein initially opposed the idea! After reading Minkowski's celebrated 1908 lecture in which he introduced spacetime, Einstein dismissed it as "superfluous erudition," muttering, "No statement can be more banal than that our familar world is a four-dimensional space-time continuum." Yet Minkowski's formulation of relativity proved indispensable, and eventually Einstein conceded that without those ideas "the general theory of relativity might have been stuck in its diapers." It is through Minkowski's formulation that students learn relativity today.

Was relativity Einstein's creation? I think so, most physicists think so, but for a century now partisans have bared fang against claw about the extent to which Einstein based his work on that of others with little or no acknowledgment. As ever more documents come to light, it does appear that Einstein could have been more generous to his forebears, in particular to Henri Poincaré.

To understand the genesis of relativity we first return yet again to the Panopticon's exhibit on the ether. We recall from the previous chapter Heinrich Hertz's cry of exaltation, "Take the luminiferous ether out of the world, and electric and magnetic forces can no longer travel through space." Despite such enthusiasm and the myriad theories to go with it, scientists had absolutely failed to imagine a set of plausible properties to ascribe to the ether.

No one could detect it either. Imagine water waves rippling along in a big aquarium, which is itself rolling down a railway track. If the waves are moving in the direction of the tank, the velocity of the waves with respect to the ground should be simply the velocity of the waves in the tank *plus* the velocity of the aquarium itself, as you would expect. If the water is rippling oppositely to the aquarium's motion, then of course the velocity of the waves relative to the ground should be the velocity of the waves in the tank *minus* the tank's velocity. The same applies to light waves rippling over, say, the Earth as it moves through absolute space—the ether. The waves' velocity relative to the ether (ground) should be the sum or difference of the velocity of light with respect to the Earth (tank) and the Earth's own velocity with respect to the ether. If you further imagine bouncing a series of water waves in the tank off one wall, the crests and troughs of the reflected waves will set up an interference pattern with crests and troughs of the incoming waves, exactly as Thomas Young would have predicted. But in a moving tank, the reflected waves will be traveling at a different velocity (relative to the ground) than the incoming waves, and hence the locations where crests and troughs line up will be shifted. The interference pattern will be displaced. So it is with light.

However, a number of experiments, the most famous being the Michelson-Morley experiment of 1888, failed to detect any such "ether drift." In his 1905 paper Einstein makes only a famously vague reference to "unsuccessful attempts to detect motion with respect to the 'light medium,'" and so it has generally been assumed that the Michelson-Morley experiment had little impact on his thinking. Little impact, maybe, but the man knew about it. In a letter to Mileva in 1899 Einstein mentions that the previous year he had read a "very interesting paper" by the physicist Wilhelm Wien. He even wrote to Wien about it. Wien's paper lists thirteen experiments made to detect motion through the ether, including the Michelson-Morley experiment. In 1895 Einstein had also studied a paper by Hendrik Lorentz, which is all about the Michelson-Morley experiment and in which Lorentz discusses his famous "Lorentz contraction," describing how objects might shrink as they moved. Lorentz's proposal, first made in 1892, was totally ad hoc; he introduced it soley to explain away the null results of the Michelson-Morley experiment and to save the ether.

Lorentz, by the way, was only the third person to posit the "Lorentz transformation" equations that produce the celebrated contraction by mixing up space and time in the way Einstein would eventually explain. As early as 1887 Woldemar Voigt showed that an equation of this type had the properties that Lorentz would need. (Voigt was working purely mathematically; he did not actually contemplate the motion of physical objects.) Although Lorentz and Voigt were acquaintances, Lorentz, strangely, does not appear to have immediately known of Voigt's work. When he found out about it he readily conceded he could have borrowed the result. Two years after Voigt, in 1889, the brilliant Irish physicist George Fitzgerald became the first to propose that material bodies might contract when they move. Fitzgerald and Lorentz were also friends, but again, Lorentz missed Fitzgerald's contribution until after 1892. From then on Lorentz always gave Fitzgerald credit, and for this generosity, a century later the hyphens following Lorentz's name (Lorentz-Fitzgerald-Voigt) have vanished.

There were a number of other major problems regarding light and the ether that led to relativity. We need to mention two. The first was the aberration of starlight. Imagine yourself in a rainstorm. Barring wind, the rain falls straight down. But if you are in a moving car, the rain appears to strike the windshield at an angle, due to the fact that the raindrops have now picked up a horizontal velocity relative to the car. The same pertains to starlight. Earth is moving with respect to the ether. As a result, starlight that was originally streaming straight down is now deflected (aberrated) by a certain angle and so, to an earthbound telescope, the star appears to be in the wrong position. The amount of aberration of starlight is actually fairly large—large enough to have been observed by James Bradley by 1729. However, according to some theories, the ether should move along with the Earth, in which case aberration should be zero. To give a proper explanation of aberration became a major impetus for Einstein's development of relativity.

But perhaps the most important fact that entered into Einstein's thinking was the peculiarity of Maxwell's equations of electricity and magnetism. When James Clerk Maxwell finished his theory of the electromagnetic field in 1865 he noted something very strange about his results: Light waves appeared to travel at a constant velocity. There was no mention of any particular system, moving or otherwise, in his equations. The velocity of light neither increased nor decreased, whether it was emitted on the ground or from a moving train. This peculiarity struck Maxwell hard; he began to think either that his equations were wrong or that they were valid only in a system at rest with respect to the ether.

Einstein took the bull by the horns: He decided that Maxwell's equations were correct in any reference frame. Regardless of the system's velocity, Maxwell's equations *did not change*. They were *invariant;* the speed of light *was* constant. What was wrong, Einstein decided, was the entire concept of the ether. Michelson and Morely couldn't measure the velocity of the Earth with respect to this absolute standard of rest because this absolute standard of rest didn't exist. As a consequence, velocities could not get added and subtracted in the simple way we did above, which assumes absolute rest. And with that, the entire edifice of Newtonian physics collapsed. From the ashes sprang special relativity, in which velocities got added in strange ways, but in which the velocity of light was always a constant and in which the notions of absolute space and absolute time vanished forever.

Or did they?

There can be no doubt that Einstein thought long and hard about relative motion. Historian of science Jagdish Mehra has discovered a short essay Einstein sent to his uncle at the age of fifteen or sixteen in which he describes his first "modest expression of some simple thoughts" on how to investigate the properties of the ether with magnetic fields. A few years later, in an outpouring of tenderness to Mileva, he wrote, "I am more and more convinced that the electrodynamics of moving bodies, as presented today, is not correct and that it should be possible to present it in a simpler way. The introduction of the term 'ether' into theories of electricity has led to the notion of a medium of whose motion one can speak without being able, I believe, to associate a physical meaning to it." As early as December 1901, the young lover writes, "I am now working eagerly on an electrodynamics of moving bodies, which promises to become a capital paper. I wrote to you that I doubted the correctness of the ideas about relative motion. But my doubts were based solely on a simple mathematical error. Now I believe in it more than ever. How is your little neck?" (What became of this "capital paper" is unknown.)

Although we are astounded at Einstein's precocious ideas about relative motion, others were also thinking along similar lines, in particular Hendrik Lorentz and Henri Poincaré. We've already mentioned that Einstein read Lorentz's 1895 paper about the Michelson-Morley experiment. In 1904 Lorentz published another paper in which he showed that Maxwell's equations behaved correctly (were invariant) under the Lorentz transformation. Historians usually state that Einstein wasn't aware of this important paper, but it seems unlikely. First, Lorentz published it in the *Annalen der Physik,* the very journal in which Einstein would later publish his own theory—and where he had *already* published several papers since 1901. Second, there is a phrase in Einstein's paper in which he actually states that "Lorentz's theory . . . is in agreement with the principle of relativity." The only version of Lorentz's theory in accord with Einstein's own is the 1904 version.

And Poincaré? History has not been kind to the master's contributions. But in his *Science and Hypothesis,* first published in 1902, Poincaré boldly declares:

1. There is no absolute space, and we only conceive of relative motion; and yet in most cases mechanical facts are enunciated as if there is an absolute space to which they can be referred.
2. There is no absolute time. When we say that two periods are equal, the statement has no meaning and can only acquire a meaning by a convention.
3. Not only have we no direct intuition of the equality of two periods, but we have not even direct intuition of the simultaneity of two events occurring in two different places.
4. Finally, is not our Euclidean geometry in itself only a kind of convention of language?

These ideas are at the heart of relativity, and it is difficult to believe they did not have a profound effect upon Einstein's thinking. Poincaré was also the first to use the term "principle of relativity," which is also stated forthrightly in *Science and Hypothesis.* In a famous 1904 speech at the International Congress of Arts and Sciences in St. Louis, Poincaré even glimpses a new theory in which "the velocity of light becomes an impassable limit." But the mathematician did more than make oracular pronouncements; he wrote a pair of technical papers on Lorentz's theory, and in the longer one, completed just before Einstein's own, he has nearly everything his shadowy rival does, and in some respects more. In that paper, Poincaré shows, as Lorentz did, that Maxwell's equations are invariant if the Lorentz transformation is correct; he anticipates Minkowski's combining of space and time, and he virtually derives $E = mc^2$. What Einstein did in those fateful weeks that Poincaré did not was to show that the whole thing results from just the two postulates: the principle of relativity and the constancy of the speed of light. There is no reason to believe that the two knew of each other's work as the summer of 1905 sped toward them.*

* George Sudarshan (chapter 10) insists not only that Einstein must have known about Bachelier's work but that documentary evidence exists proving that around 1905 Einstein attended a seminar at University of Bern dedicated to Poincaré's theory of relativity, and that thus Einstein knew about it. If true, this would be an important development in the history of science, and so I put two colleagues at Bern, Hans Bebie and Viktor Gorgé, on the hunt. However, after a fairly thorough search, they turned up nothing.

Why did Einstein beat Poincaré in the race for relativity? I think the answer is that Poincaré was smarter than Einstein. Poincaré was a universalist who had, by all accounts, the swiftest mind of his day. All science came within his purview, and he flitted as a hummingbird from one occupation to another. I imagine he was easily bored. There is much to be said for being a plodder.

So I think it is fair, despite the detractors, to award creation of relativity primarily to Einstein, but it does Einstein no credit that for fifty years he never mentioned Poincaré's contributions.

It is also fair to say that Einstein did not entirely understand his own theory. And since no one bothers to read the original papers anymore (all physicists *know* relativity), no one recognizes the rather large blunder in Einstein's exposition. This is not to say that the theory is wrong, just that Einstein didn't interpret it entirely correctly. As is often the case, the equations were smarter than the author.

Einstein's error concerned nothing other than stellar aberration, one of the principal phenomena he devised the theory to explain. As mentioned, aberration is the deflection of starlight from its true path due to the motion of the Earth. Or is it the motion of the star? Or the motion of both? In his derivation of the aberration formula, Einstein clearly assumes that the velocity involved is the relative velocity between the Earth and the star. This might seem entirely reasonable in a theory that banishes absolute velocities, but, strangely, it isn't true. Imagine two stars in the exactly same position (a physical impossibility, of course, though a useful picture), one moving and one not. They each give off a photon. Once emitted, the two particles will both travel along identical straight lines to a telescope on the Earth, regardless of whether the source star is moving or not. Thus any deflection angle cannot depend on the velocity of the star, and so the relative velocity of the star and the Earth cannot be the right one to use. Which is the correct velocity? It turns out that the velocity involved is actually the relative velocity between the telescope itself at one instant and at a later instant.

The true explanation of aberration has been recognized by isolated physicists over the decades, and with two coauthors I once wrote a paper on the subject. The referee rejected it on the grounds that "the answer is obvious if you think about it for thirty seconds." As a result, the paper was never published, but if the referee was correct, then he should explain why textbook authors, including one of the twentieth century's most illustrious physicists, Wolfgang Pauli, have repeated Einstein's error ever since.

It is also not entirely true—despite Einstein's intent, so forcefully expressed in his letters to Mileva—that special relativity abolished the concept of the ether (a declaration also repeated in every textbook). At first Einstein thought that his theory had banished the mysterious substance to the dustbin of history. Later he changed his mind. In a rarely cited lecture given at the University of Leyden in 1920, five years after the completion of general relativity, Einstein concedes, "More careful reflection teaches us, however, that the special theory of relativity does not compel us to deny ether." What special relativity did was to abolish the notion of an ether that had a definite set of properties associated with it, for instance, the extraordinary stiffness mentioned in chapter 2. "There is a weighty argument to be adduced in favor of the ether hypothesis," Einstein goes on. "To deny the ether is ultimately to assume that empty space has no physical qualities whatever. The fundamental facts of mechanics do not harmonize with this view."

Einstein is referring to the gravitational field. With his general theory of relativity he showed, contrary to his own expectations, that even space completely empty of all matter can have dynamic properties, meaning it can affect the way objects move. Very weird. Very true. And so, after abolishing the ether with special relativity, Einstein resurrected it with general relativity, in the guise of the gravitational field itself. But gravity is a story we leave to the next chapter.

We have not forgotten $E = mc^2$. Einstein submitted his fifth miraculous paper to the *Annalen* in September of 1905. As mentioned, Poincaré also came very close to deriving it, and in isolated instances the equivalence of mass and energy had been known for some time, but Einstein seems to be the first to have postulated it as a general law of nature. Well . . .

What is not usually acknowledged is that in his paper Einstein only derives the famous formula in an approximate way. He then *assumes* it is true in general, but never proves it. (As an amusing aside, he also used the "wrong" letter: $L = mc^2$.) It seems to have been Lorentz, in 1911, who first showed that the formula was really true.

As a less amusing aside, somehow the idea has arisen that this most famous of all equations paved the way for the creation of the atomic bomb. It is simply not true. In 1903, two years before Einstein derived his equation, Frederick Soddy and Ernest Rutherford discovered the transmutation of elements through radioactive decay and measured the energy released. They were astounded. "The energy of radioactive change," they wrote, "may be a million times as great as the energy of any molecular change." "If it could be tapped," Soddy prophetically declared just one year later, "what an agent it would be in shaping the world's destiny. A man who put his hand on the lever . . . would posses a weapon by which he could destroy the earth." Einstein's formula allowed scientists to calculate from basic principles the amount of energy that might be released in a bomb, but that one could be built without $E = mc^2$ is not in doubt.

So there we are. Once Einstein finished his theory, it is said that he was speechless with happiness, and to one colleague he exclaimed, "My joy is indescribable." Einstein's bouts with supreme joy would become frequent. In any case, he did create special relativity, with a bit of uncredited help from associate producers, and since then many nonsensical things have been said about it. As to why Einstein was so reticent about acknowledging others, this is more a matter of psychology than of history. "Do you really believe that you could find permanent happiness through others, even if this be the one and only beloved man?" he once wrote to a female acquaintance. "I know this sort of animal personally, from my own experience as I am one of them myself. Not too much should be expected from them, this I know quite exactly. Today we are sullen, tomorrow high-spirited, after tomorrow cold, then again irritated and half-sick of life—and so it goes—but I have almost forgotten the unfaithfulness & ingratitude & selfishness, things in which almost all of us do significantly better than the good girls." This self-portrait, penned in 1899, when Einstein was only twenty, is perhaps the best answer to the question, and it would remain a good answer during the years of developing his supreme creation—the general theory of relativity.

8 / What Did the Eclipse Expedition Really Show? And Other Tales of General Relativity

"One of the great contents of my life," the philosopher Paul Oppenheim once recounted to me, "is that I was a friend of Albert Einstein. We lived near each other in Berlin. I remember, when came in the results of the eclipse expedition and confirmed his theory, I ran down the street to his house and great with agitation shouted, 'Einstein! Einstein! Your theory has been proven!' He looked at me and answered calmly, 'It would have been too bad for God had I been wrong.'" The London *Times* agreed, less calmly:

REVOLUTION IN SCIENCE

NEW THEORY OF THE UNIVERSE

NEWTONIAN IDEAS OVERTHROWN

Space Warped

The *New York Times* agreed:

ALL LIGHTS ASKEW IN THE HEAVENS

STARS NOT WHERE THEY SEEMED OR WERE CALCULATED TO BE,

BUT NOBODY NEED WORRY

A Book for 12 Wise Men

The *Berliner Illustrirte* agreed: "A great new figure in world history: Albert Einstein, whose investigations signify a complete revision of our concepts of Nature, and are on a par with the insights of a Copernicus, a Kepler, a Newton." The chair of the Royal Society, J. J. Thomson, agreed: "This is the most important result obtained in connection with the theory of gravitation since Newton's day."

The most important result was the eclipse expedition of 1919, led by Sir Arthur Eddington. It is the most famous eclipse expedition that was, that is, and that shall ever be. It is the expedition that confirmed Einstein's general relativity, his prediction that light itself should be deflected by the gravitational field of the sun, that the images of stars near the sun

77

should be shifted by 1.74 seconds of arc. This minute angle, the angle made by a dime at a distance of two kilometers, changed the shape of the universe forever.

Did the eclipse expedition measure it?

It was in an Oxford pub one night (actually, the astrophysics department—which, however, is not far from the nearest pub) that Dennis Sciama, a leading cosmologist who had been graduate advisor for Stephen Hawking, among others, sallied forth: "If you imagine that the answer could have been anything between zero and infinity, and you ask yourself what the odds were that Eddington would have gotten 1.74 seconds of arc, then you realize he knew what he was looking for."

Sciama was correct.

But in the Contemporary Panopticon of Present and Past Concepts, before one reaches the exhibit on Treatment of Data in Celebrated Experiments, one must peruse the exhibit on general relativity. The display informs us that, like "special relativity," "general relativity" is a misnomer. General relativity is more properly known as Einstein's theory of gravitation. There is no doubt that he created it. Einstein began thinking about general relativity shortly after completing special relativity. Special relativity dealt only with objects moving in uniform motion—constant velocity. (Hence "special.") Einstein wanted to incorporate changes in velocity—acceleration. (Hence "general.") The breakthrough came in 1907 when he suddenly had "the happiest thought of my life." Happy thoughts occurred to Einstein often. Still, his thought of 1907 (presumably he had more than one) is undeniably a good candidate for supremacy, and at the Panopticon it is centrally located. When you jump off the roof of a house you are accelerated downward by the earth's gravitational field. Nevertheless, for the brief instant before you hit the ground, you feel weightless. It is easy to verify this experimentally: Hold an apple while you jump, and you will see that it falls as the same rate as you do. Jump while standing on a bathroom scale; it also falls at the same rate as yourself. Consequently you are not pressing down on the scale, and because nothing is pressing on it, it reads zero—you are weightless. Einstein realized that the downward acceleration is exactly canceling out the gravitational attraction of the Earth. Conversely, a person accelerating upward in an elevator or a rocket ship feels heavier than normal. Thus the Principle of Equivalence: Acceleration and a gravitational field are indistinguishable. If Einstein wanted to create a theory of relativity that incorporated accelerations, he was going to have to create a new theory of gravity.

Einstein worked on his theory for the next eight years. By 1911 he concluded that light itself should be deflected in a gravitational field, and went on to calculate that a light ray passing near the sun should be deflected by an angle of about .83 seconds of arc (more accurately .87). For this calculation he had to make certain assumptions: roughly speaking, that light behaved according to ordinary Newtonian physics. He ended the paper with a plea: "It is urgently desirable that astronomers concern themselves with the question brought up here, even if the foregoing considerations might seem insufficiently founded or even adventurous."

Actually, Einstein's 1911 answer was first achieved in 1801 by the German astronomer Johann Georg von Soldner, who also treated light as a Newtonian particle. In 1783 John Michell had presented a paper before the Royal Society in which he concluded that the gravitational attraction of a star about five hundred times more massive than the sun would be so great that even light could not escape, a concept that may sound familiar. In 1796 Pierre-Simon de Laplace independently came to the conclusion that only about 250 solar masses were necessary for a star to become invisible. Henry Cavendish evidently read Michell's paper, and Soldner evidently read Laplace's. Cavendish made some calculations for his notebook, while Soldner published a paper so far ahead of its time that it was invisible to his colleagues: "Hopefully, no one would find it objectionable that I treat a light ray as a heavy body," he wrote. No one in the twentieth century, perhaps. The dawn of the nineteenth? Treating light very much as Einstein did over a hundred years later, he calculated almost exactly the same deflection: .84 seconds of arc. Soldner's paper was retrieved from oblivion only in 1921, when the virulent anti-Semite Philipp Lenard republished it to discredit Einstein. Its prescience stuns even today. Soldner's work shows it is not always necessary to be Einstein to be Einstein and that simple considerations are sufficient to get approximate answers.

Approximations. It is well known among physicists that Einstein's 1911 value gave the wrong answer, specifically half the right answer. Later Einstein expressed relief that no eclipse expedition had been sent out at the time. (There were positive aspects to World War I.) What is less well known is that his second derivation, done in 1915 when the general theory was complete, was also incorrect! The answer happened to be right, but the method was wrong.

General relativity is based on two principles: the equivalence of gravity and acceleration and, to the left of it in the Panopticon, that the results of a theory should not depend on the coordinate system one uses. What does this mean? Most people are familiar with ordinary rectangular coordinates, x, y, and z, but physicists and mathematicians use many other systems, depending on the problem. There may be curved grids, there may be falling grids, there may be spinning grids. All of these are mere methods of description, not reality. Reality should be independent of coordinates. This is one of Einstein's most beloved principles, the Principle of General Covariance.

Unfortunately, he seems to have forgotten it when he derived light deflection by the sun. His derivation definitely employs quantities that are meaningless in the context of relativity. Basically, for the velocity of light he takes the distance light travels in rectangular coordinates divided by time. This sounds reasonable. But in another system the number could be entirely different. It is somewhat as if he said, "The velocity of light is ³/₄." Until you attach some physical units to ³/₄, like "³/₄ furlong per fortnight," the statement is devoid of content.

How did Einstein manage to get the correct answer by the wrong method? The Panopticon is filled with such exhibits; they occur frequently, especially when people like Einstein are making the mistakes. Einstein's nose for the right answer was unerring; mathematics was a distraction. In this case, the amount of spacetime curvature causing the deflection was so small that the coordinate system Einstein did employ was adequate.

A proper derivation of light deflection requires consideration of how light rays travel in the gravitational field of a star. On December 22, 1915, only one month after general relativity was completed, Karl Schwarzschild wrote to Einstein from the Russian front, where he was stationed calculating artillery trajectories. Apparently military mathematics did not sufficiently tax Schwarzschild, for the letter contained the results of a "not overly long calculation" employing Einstein's theory that gave the exact gravitational field around any spherical object. The Schwarzschild solution was the first relativistic description of black holes. The usual story is that Einstein wrote back dramatically, saying, "I would not have thought an exact solution was possible," a letter Schwarzschild got on his deathbed. What Einstein really said was, "I would not have thought that the strict treatment of the [mass]-point problem was so simple." That Schwarzschild received Einstein's letter on his deathbed seems unlikely; he had already returned to Berlin in March, seriously ill with a rare skin disease. On May 11, 1916, he died. Schwarzschild's result is considered the greatest discovery in pure general relativity of the twentieth century, after the creation of the theory itself.

Einstein read Schwarzschild's paper before the Prussian Academy of Sciences on January 13, 1916. Schwarzschild might well have used his solution to calculate light deflection, but he did not. The first proper calculation was not carried out until the 1920s, long after the celebrated eclipse expedition. What is not generally known, even among specialists, is that on May 27, 1916, Hendrik Lorentz himself presented to the Royal Dutch Academy a remarkable paper by one of his students, Johannes Droste. It contained the same solution for the gravitational field of a spherical star that Schwarzschild had achieved in December, but obtained more clearly; it contained a derivation of the perihelion shift of Mercury from first principles, as Schwarzschild had done (more on perihelions momentarily); and it contained a detailed analysis of all possible orbits in this gravitational field, conclusions that anticipated certain results of black-hole physics discovered only decades later.

Schwarzschild's paper achieved immortality, Droste's oblivion. Why? Schwarzschild was a few months ahead of Droste; that cannot be denied. Einstein's advocacy of Schwarzschild could not have hurt; neither did Schwarzschild's visible position as director of the Astrophysical Observatory. Nevertheless, by almost any measure other than the date, Droste's paper was superior to Schwarzschild's, and his cruel fate seems contrary to the usual overgenerosity of relativists in honoring independent discoveries. For instance, the standard big-bang cosmology is named after four people who discovered and rediscovered it over a period of thirteen years. An insignificant four months separates Droste's discovery from Schwarzschild's. It would be appropriate to begin referring to the Schwarzschild-Droste black hole.

Behind the perihelion shift of Mercury, first approximated by Einstein, then derived rigorously by Schwarzschild and Droste, lies another tale. A lone planet governed by strict Newtonian physics travels along an elliptical orbit, and the point of closest approach to the Sun—the perihelion— always remains at the same point in space. Due to the gravitational attraction of the other planets, however, the position of the perihelion will advance with time, producing a sort of petal pattern. At the tail end of chapter 4, Urbain Le Verrier had discovered that when the gravitational influence of all the known planets was subtracted, a mysterious surplus of 38 arc-seconds per century (later amended by Simon Newcomb to 43 seconds) remained for Mercury's orbit. All attempts to explain it within the context of Newtonian physics failed.

As Einstein finished general relativity in November 1915, he was able to show that his theory, by modifying Newtonian gravity, predicted the perihelion shift exactly. "Imagine my delight at realizing . . . that the equations yield Mercury's perihelion motion correctly. I was beside myself with joy and excitement for days," he wrote to Paul Ehrenfest. He told other friends that his discovery had given him palpitations of the heart and yet others that when he saw that his calculations agreed with the observations he had the feeling that something inside him had burst. Most biographers describe this moment as the most intense of Einstein's life (another candidate for supremacy).

We can appreciate Einstein's excitement on placing the capstone on general relativity, perhaps the greatest single intellectual achievement in history. But as the story is usually told—or at least understood—the perihelion calculation popped out at the last moment with little forethought, a bonus. This impression is furthered by Einstein's own remarks: The theory "explains quantitatively the rotation of the orbit of Mercury, discovered by Le Verrier . . . without the need of any special hypotheses." One is tempted to say, *Come on, Al.* Einstein had written to Conrad Habicht about the perihelion shift of Mercury as early as 1907 and had been after it ever since. Mercury was one of the reasons Einstein created general relativity in the first place, and in 1912 he and his friend Michele Besso discarded an entire version of the theory because it got the perihelion shift wrong. The manuscript, in German, is available in the Princeton University Press collection. To be sure, the final version of general relativity explained Mercury's orbit without the need for "special hypotheses," but that was after all the previous versions had been thrown out.

Completing general relativity was "the greatest satisfaction of my life." (Has he exceeded his quota of superlatives?) In fact, Einstein was not even the first person to write down the final equations of his own theory. The mathematician David Hilbert beat him by five days, but in this case Einstein had been publishing preliminary papers for several years and had briefed Hilbert on his progress in a number of lectures. There is no question that general relativity belonged to Einstein.

World War I was out of the way, and eclipse expeditions could proceed. Einstein's home life at the time must have been confusing, if not bizarre. As I've mentioned, the world knows by now that his relationships with women were far from saintly. The Princeton collection indeed reveals libertinistic tendencies. Ilse, his second wife Else's daughter from a previous marriage, writes to a friend, Georg Nicolai, on May 22, 1918, detailing an extraordinary situation:

Dear Professor,

You are the only person to whom I can entrust the following and the only one who can give me advice, and that is why I ask you please to consider carefully what I am writing you now, and then let me know your view. You remember that we recently spoke about Albert's and Mama's marriage and you said to me that you thought a marriage between Albert and *me* would be more proper. I never thought seriously about it until yesterday. Yesterday, the question was suddenly raised about whether A. wished to marry Mama or me. This question, initially posed half in jest, became within a few minutes a serious matter which must now be considered and discussed fully and completely. Albert himself is refusing to take any decision, he is prepared to marry either me or Mama. I know that A. loves me very much, perhaps more than any other man ever will, he also told me so himself yesterday.

Ilse goes on to explain that she loves A. very much as a person, although she has no desire to be physically intimate with him. A., on the other hand, told her "how difficult it is for him to keep himself in check." Mama, for her part, appears willing to step aside should A. decide to marry I. I. feels she ought not to marry Albert but seeks Nicolai's advice. She requests, "Please destroy this letter immediately after reading it." A. and E. are married on June 2, 1919, four days after the eclipse that will canonize him.

The eclipse was not the first eclipse. The German astronomer Erwin Freundlich read Einstein's (incorrect) 1911 paper on light deflection, took up the cause, and organized a multinational expedition to Brazil to observe the eclipse of October 10, 1912. It rained. One astronomer reported, "We suffered an eclipse instead of observing one." Undeterred, Freundlich organized a second expedition, to Russia for the eclipse of August 21, 1914. Einstein was in on it; he helped raise funds, which were provided by the chemist Emil Fischer and by Krupp, the arms manufacturer.

They say eclipses are evil omens. On July 28 a Serbian nationalist assassinated the Archduke Ferdinand, and on August 1 Germany declared war on Russia. Freundlich, on Russian territory, was arrested as an enemy alien and interned with his equipment. After a month he was exchanged for Russians in parallel circumstances. The Germans were not alone in Russia. The director of the Lick Observatory, William Campbell, had also organized an expedition to measure light deflection. Bad weather did them in; they returned to California without starlight and without equipment. If they had not been forced to store their telescopes in Petrograd for the duration of the war, history might have been written otherwise.

In 1918 an eclipse passed over the United States. Circumstances were ideal. The Lick astronomers hardly had to do more than walk outside to observe it. Unfortunately, they had nothing to observe it with. Their equipment, shipped with foresight a year earlier from Petrograd, was stalled in Japan due to the war. They improvised with borrowed lenses and on June 8, 1918, captured the first pictures of deflected starlight. Again the war intervened. Heber D. Curtis, in charge of the photography, was on military leave during the eclipse but then returned to service, and so nothing was done with the plates for ten months. Eventually they were developed. Analysis of the data, reanalysis, and arguments between Campbell and Curtis dragged on for several years. At one point Campbell wrote to Curtis that the latter had made so many arithmetical mistakes that the final results were no better than the originals—both were subject to the laws of random errors.

Despite the arguments and delays, Campbell did report to the Royal Society of London in early 1919 that in his opinion the Lick results "preclude the larger Einstein effect, but not the small amount expected according to the original Einstein hypothesis." In other words, the correct answer of 1915, arrived at by incorrect means, was wrong, but the incorrect answer of 1911, originally derived in 1801, was possibly right.

And thus the road was cleared for Arthur Stanley Eddington.

At the close of World War I general relativity was virtually unknown in England. Eddington, secretary of the Royal Astonomical Society, was its lone champion. He had arranged for publication of a series of papers on relativity by the Dutchman Willem de Sitter and had written a review himself. These were the only papers on relativity published in England before 1919. Eddington advocated Einstein's theory to the astronomer royal, Frank Dyson, who, infected by Eddington's relativitis, proposed the expedition.

It appears that their enthusiasm contained extrascientific elements. According to Subrahmanyan Chandrasekhar, of the Chandrasekhar limit, who had been a student of Eddington's, the matter was not only relativistic but religious. Eddington, a Quaker, had refused military service during the war as a conscientious objector. Conscientious objectors were sent to internment camps. To avoid this, Dyson intervened, expressly promising the Home Office that if the war ended by 1919, Eddington would lead one of the two expeditions planned to test Einstein's theory.

In March 1919 two ships set sail. The first team, consisting of two of Eddington's colleagues, headed to Sobral, a town in northern Brazil. The second, consisting of Eddington himself and another colleague, went to the island of Principe, off the coast of Africa. What happened was this:

The Brazilian team, A. Crommelin and C. Davidson, had brought two instruments, an "astrographic" telescope and a four-inch telescope. During the eclipse they obtained eighteen usable plates with the astrographic and eight usable plates with the four-inch. Eddington and his partner, Cottingham, on Principe, had only a single telescope, and the day of the eclipse was cloudy. They took sixteen photographs, all but two of which turned out to be useless.

Photographing eclipses is more involved than taking a snapshot. To calculate light deflection, one must compare the positions of the stars during the eclipse with their positions in the same star field in the absence of the sun. This requires shooting comparison plates some months before or after the eclipse when the same stars appear in the night sky. To make matters worse, the slightest change in temperature or alignment of the telescope optics will change apparent stellar positions. To correct for this, one needs at least six stars on each plate.

Most of the astrographic plates from Brazil showed twelve stars, the four-inch plates seven, the two African plates five. Nor were Eddington and Cottingham able to remain on Principe the required months to take the comparison photographs. This was done back at Oxford. The best plates were unquestionably those from the four-inch telescope at Sobral. When Crommelin and Davidson calculated the deflection of light from these plates, they got 1.98 seconds of arc with a probable error of .12 seconds.* Even with the error the result was higher than Einstein's prediction. The data from the poorer eighteen astrographic plates gave a deflection of .86 seconds of arc—almost exactly Einstein's 1911 value.

Eddington performed a unique data analysis. As far as one can tell from the involved procedure, it assumed Einstein was correct to begin with. He obtained for his two plates 1.61 seconds of arc, with a probable error of .3 seconds.† At the upper end this result included Einstein's prediction; at the lower end, it was just about as close to the 1911 "Newtonian" prediction. If one kept all the data, about the only thing a reasonable

---

* Probable error is an obsolete statistical measure. In terms of standard deviations, the Sobral 4″ results were .178″, the Sobral astrographic .48″, and the Principe astrographic .444″.

† The number ±.3 seconds comes from using check plates. Apart from the comparison plates of the star field that must be taken months later, astronomers also take check plates of arbitrary fields of stars the night before or after the eclipse, both in England and at the eclipse site. This is to make sure that temperature changes or transportation have not altered the alignment of the optics or altered the focus of the telescope. If something is out of whack, then the stellar images taken at the eclipse site will be displaced relative to the same images taken in England. If all is well, the check plates determine the scale of image displacement on the eclipse photographs.

person would conclude was that the gravitational deflection of light by the sun was somewhere between Einstein's mistaken value of 1911 and the "correct" prediction of 1915. But history is not made by reasonable people.

November 6, 1919, the extraordinary joint session of the Royal Society and the Royal Astronomical Society. You are there. Sir J. J. Thomson, first, second, or third discoverer of the electron, now president of the Royal Society, is in the chair. After stating that the uncorrected results give values of .97 seconds and 1.40 seconds, Thomson asserts that the "much better plates" give 1.98 seconds. Then he declares, "After a careful study of the plates, I am prepared to say that there can be no doubt that they confirm Einstein's prediction. A very definite result has been obtained that light is deflected in accordance with Einstein's law of gravitation."

A murmur of tacit approval, we can be sure, runs through the audience. Also running through the audience is the tacit assumption that three results and three results alone are possible: no deflection, the Newtonian value of .87 seconds, and the full Einsteinian value. No other theories are allowed.

Dyson mounts the podium. He describes the Principe expedition but inexplicably fails to mention the bad data it obtained. Suffering a second memory lapse, he also fails to mention the astrographic results from Sobral, which give .86 seconds. Of the three sets of data, only the four-inch results remain: 1.98 seconds of arc. The other two are consigned forever to oblivion.

Eddington now speaks. He also mentions the four-inch results and his own Principe results, which "support the figures obtained from Sobral." As Campbell himself remarked a few years later, "Professor Eddington was inclined to assign considerable weight to the African [Principe] determination, but, as the few images on his small number of astrographic plates were not so good as those on the astrographic plates secured in Brazil, and the results from the latter were given almost negligible weight, the logic of the situation does not seem entirely clear."

To be sure, in Eddington's later and excellent popular writings, he spends the bulk of the narrative on the African expedition, manages to "condemn" (his own words) the Sobral astrographic results, but bravely stands a single African plate in support of the four-inch results. He also admits to being "not altogether unbiased."

But on that historic day in November 1919 no voices are raised. Then the astronomers go out to talk to the journalists.

*   *   *

In the Contemporary Panopticon of Present and Past Concepts, one exhibit has only recently been restored to the Division of Natural Sciences: Personality. The eclipse data did not prove Einstein correct; Dyson and Eddington did. The concientious objector Eddington saw the expedition not only as a vindication of Einstein but as a testimony to the international character of science, a rapprochement with Germany after the war. To be sure, he confided to Chandrasekhar that, had it been up to him, he would not have undertaken the expedition at all; he was already convinced of the truth of Einstein's theory.

What of Einstein? "It would have been too bad for God had I been wrong," he told Oppenheim. The story is so widespread that we can assume Einstein made the remark on many occasions. One suspects he polished the line until it became the happiest of his life. Einstein's confidence is frequently on display. In 1914, with only the mistaken 1911 calculation in hand, he told Michele Besso, "I do not doubt anymore the correctness of the whole system, whether the observation of the solar eclipse succeeds or not."

The restored exhibit in the Panopticon informs us that in the eighteenth century personality was considered external; one put it on like a newly tailored coat. Einstein belonged in the eighteenth century. He owned several coats. At times he exchanged the coat of confidence for others, less lustrous, hidden in the Panopticon's storeroom. He participates in the planning and funding of the previous eclipse expeditions. To Arnold Sommerfeld in 1915 he reports Freundlich's failed attempts to measure light deflection, adds he is unconcerned because "the theory seems to me to be adequately secure," then signs the letter, "Your raving Einstein." To Ehrenfest he inquires with deliberate casualness in September 1919, "Have you heard anything about the English solar eclipse expedition?"

Einstein was canonized. His theory of general relativity is about as correct as a scientific theory gets, but it took another fifty years before eclipse measurements became accurate enough to show it. Eddington's original photographic plates undoubtedly still reside at the Royal Astronomical Society. Someone might reanalyze them. Only one result is sure: The tale that the 1919 eclipse expedition verified Einstein's theory will forever remain untarnished.

9 / Two Quantum Tales: Bohr and Hydrogen, Dirac and the Positron

I

Quantum mechanics, with relativity the great theory of the twentieth century, is also the most mysterious. It came into being in 1900 when Max Planck found that in order to explain the spectrum of radiation given off by hot bodies, he was forced to assume that the energy they emitted could not take on any value but came only in discrete units—in his term, quanta. Planck's idea was so revolutionary (if any breakthrough in the Panopticon of Present and Past Concepts deserves the term, it is this one) that even he could not accept it, let alone anyone else. Five years would pass before Albert Einstein wielded Planck's hypothesis to explain the photoelectric effect, and in so doing achieved quantum mechanics' second spectacular triumph. Quantum's third great victory took place in 1913 when Niels Bohr used Planck's ideas to devise a model of the hydrogen atom that for the first time was capable of explaining the hydrogen spectrum.

Now, try the following experiment: Walk into the nearest physics department, casually mention that the Bohr atom was based on two crucial assumptions that flew in the face of classical atomic theory, and ask the nearest physicist what those assumptions were and which was the most important. Your victim may at first not understand the question, and you may have to repeat yourself while he scratches his head and pulls the nearest textbook off the shelf. But once he's with you, with a probability of five out of six he will answer, "Of course. The important one is the quantization of the electron's angular momentum." You may not quite understand what this means, but he won't notice. To be sure, if he has happened to pull down the widely used *Modern Physics* by Serway, Moses, and Moyer, he may emphasize the miraculousness of the achievement by quoting, "The concept of angular momentum quantization seems to have sprung full blown from Bohr's *Gedankenkuche* (thought kitchen)."

Whether the quantization of the electron's angular momentum was the key ingredient of the Bohr atom has been debated. What can't be debated is that this ingredient was cooked up in the thought kitchen of another, forgotten chef.

After Planck had explained the radiation given off by hot objects (black-bodies) and Einstein took care of the photoelectric effect, physicists still faced what had been the third, and perhaps greatest, mystery of late-nineteenth-century physics: atomic spectra, in particular the hydrogen spectrum. When heated or excited by electricity, atoms emit light. The yellow of sodium vapor lamps, the greenish white of mercury lamps, the pinkish red of neon signs, the blinding and dangerous blue-white of halogen headlights—when analyzed through a spectrometer, each of these colors is seen to result from one or more bright lines given off by the elements at specific frequencies. Such spectral lines are the most direct and powerful key to the inner workings of atoms. In the nineteenth century, of course, this wasn't clear; it wasn't even conceivable. The very notion of atoms was still debated, and no one had any explanation of what caused the elements to produce the colors they did.

This was not for lack of trying. The most celebrated attempt to account for the hydrogen spectrum was made in 1885 by a sixty-year-old Swiss schoolteacher named Jacob Balmer, who happened to be a numerologist. His earlier habilitation thesis had been entitled "The Prophet Ezekiel's Vision of the Temple Broadly Described and Architechtonically Explained," and in it Balmer disclosed the secret numerical relationships embodied in the Temple of Solomon, just as his esteemed colleagues did with the Great Pyramid and as his descendants today tackle the numerical mysteries of the Washington Monument. Balmer wrote only one other paper that anyone remembers. It was not much different in approach from his thesis, except that instead of describing the proportions of the Temple, he intended to describe the spectrum of hydrogen. By trial and error he managed to devise a simple numerical formula that gave the frequencies of the hydrogen spectral lines *exactly*. Balmer had no more idea of why his formula worked than he did for the proportions of the Temple, unless it had something to do with Ezekiel, and to explain it became a major industry for the next twenty-five years.

Finally, in 1913, Niels Bohr took the prize. "As soon as I saw Balmer's formula, the whole thing was immediately clear to me," he said more than once. It's an explanation to satisfy the popular imagination; certainly it satisfies the authors of *Modern Physics*. But what had prepared Bohr's mind so that a single glance at the Balmer formula was enough?

* * *

Niels Bohr is often considered the greatest physicist of the twentieth century after Einstein, and he is second to none in the respect and affection he commands from the elder generation of physicists. As one prominent physicist who knew both Bohr and Einstein said to me not long ago, "Bohr was ten times the human being Einstein was." He was the great father figure of physics, and it is difficult to think of anyone now alive that the current generation of scientists turns to with equal veneration.* Bohr, born in 1885, came from an extraordinary family. His father, an eminent physiologist, was nominated for a Nobel prize; Niels himself would win the prize for his explanation of the hydrogen spectrum; his brother, Harald, became as famous a mathematician as Niels a physicist; and Aga, Niels' son, also won a Nobel prize for physics.

In 1911, after receiving his doctorate in Copenhagen for research on the electron theory of metals, Niels came to Cambridge to work with J. J. Thomson. The collaboration never materialized, and the following year Bohr moved on to Manchester to work with Ernest Rutherford, who, having failed to become recognized as the inventor of the radio (chapter 14), had just announced his celebrated solar system model of the atom. This model, with electrons orbiting the nucleus like planets around the sun, is the mental picture most people conjure up to this day when attempting to visualize an atom. At the time, though, another model, Thomson's famous "plum pudding," which consisted solely of negative plum-corpuscles suspended in a uniform, positively charged pudding-background (and no nucleus!) was the favorite.† It was Bohr's good fortune to be in Rutherford's orbit at the time and his good sense to choose Rutherford's atom over Thomson's as the one most likely to be correct.

Shortly after arriving at Manchester, Bohr decided to tackle the problem of atomic spectra via Rutherford's atom. Modeling anything on the basis of Rutherford's atom and classical physics, unfortunately, brought with it several serious—even fatal—problems, one of which was that, according to Newtonian mechanics, electrons traveling in circular rings around a nucleus would begin to oscillate and eventually rip the atom apart. The Rutherford atom was unstable.‡

---

* In my own experience only John Wheeler, the guiding light of American relativity, and Dennis Sciama, the guiding light of European relativity, might come close.

† The fact that Thomson viewed an atom as composed of negative corpuscles floating in a positive sea shows that his concept of an electron was not ours (chapter 6).

‡ Another difficulty, more stressed today, is that according to Maxwellian electrodynamics, an electron in a circular orbit should radiate so much energy that it would spiral into the nucleus in less than one billionth of a second. Our continued existence suggests such a conclusion is erroneous. However, Bohr considered this problem only right at the end.

By 1912 it had become increasingly obvious that quantum mechanics was going to have to be introduced to explain atomic spectra, but no one was quite sure how to do it. According to Planck's theory, radiation from hot bodies was caused by invisible oscillators or "resonators" (the nature of which he never explained, by the way) that by oscillating emitted the quanta. But thermal radiation is found at all colors, whereas spectral lines occur only at specific frequencies, and so it was not at all obvious how to adapt Planck's ideas to this problem. As early as 1908 Johannes Stark made a suggestion relating Planck's constant to the frequency of spectral lines, and in 1910 one Arthur Eric Haas also advanced a proposal that related the famous constant to the size of an atom.* In retrospect, both ideas seem to be on the right track, but as they say, the devil lies in the details, or maybe in the Balmer formula. By the time of his doctoral thesis in 1911, Bohr was also one of those convinced of the need to break with the past—"It was in the air," he said later—but the question remained how to reproduce Balmer's results.

Bohr himself took a significant step—some would say the most significant step—in the summer of 1912 when he realized that the only way to explain the Rutherford atom's stability was to declare it so. Electrons in certain circular orbits merely orbited stably. This would become the first postulate of the Bohr atom, and he made no attempt to justify this radical hypothesis on the basis of known physics, "as it seems hopeless." But he still had no clue as to how to explain the hydrogen spectrum. Suddenly in early 1913 he produced the answer. What is striking in reading his famous paper, published that July, is the number of references not to a thought kitchen but to a rival theory by one John William Nicholson.

Erased from today's textbooks, Nicholson (1881–1955) was a prominent mathematical physicist of the early twentieth century who was born in Darlington, studied at Manchester, then moved on to various positions at Cambridge, London, and Oxford. Like Bohr, he was concerned with explaining atomic spectra—in particular those arising in astrophysical bodies—and he created the most ambitious theory of the time to do so. Nicholson postulated that the familiar elements were actually composed of "protoatoms" that might be present in the Sun or in gaseous nebulae and which he termed "coronium," "hydrogen" (not necessarily ordinary hydrogen), "nebulium," and "protofluorine." The last, for example, consisted of five electrons circling a nucleus in a Saturn-like ring, and the spectral lines

---

* Planck's constant is the number that Planck introduced to determine the magnitude of the energy quanta. The energy of any quantum is equal to Planck's constant multiplied by the frequency of the radiation.

it emitted were determined by the rate at which the electrons revolved around the nucleus and the rate at which they also vibrated perpendicularly to the ring. It was a rather complicated model, not without its problems, one of which was that there was no obvious way to fix the size of the orbits. The size would, as it turns out, determine the color of the spectral light. In his first papers, published in 1911, Nicholson has no idea how to do this, and must determine the size of nebulium by fitting it to the frequencies of known spectral lines. Nevertheless, he seems to be able to get a close match to ten real lines. Then suddenly in June 1912 he makes the following observation:

> The constant of nature in terms of which these spectra can be expressed appears to be that of Planck in his recent quantum theory of energy. It is evident that the model atoms with which we deal have many of the essential characteristics of Planck's resonators.

And a few pages later, this startling proposal:

> If, therefore, the constant h of Planck has, as Sommerfeld has suggested, an atomic significance, it may mean that the angular momentum of an atom can only rise or fall by discrete amounts when electrons leave or return. It is readily seen that this view presents less difficulty to the mind than the more usual interpretation, which is believed to involve an atomic constitution of energy itself.

Analogously to ordinary momentum, angular momentum is the measure of the oomph a body has while traveling in a circular orbit, and is merely the product of the object's mass, velocity, and orbital radius. It almost seems strange that for ten years scientists worried over the relationship of Planck's constant to the *energy* of radiation when the constant itself is in fact a unit of *angular momentum*. Nicholson was the first to perceive its significance: Planck's constant determines the orbital angular momentum of the electrons in an atom, and it can change only by discrete amounts. And having announced with crystal clarity what most physicists regard as the main assumption of the Bohr atom, he proceeds to calculate. Fixing the angular momentum fixes the allowed radii of the orbits, and so Nicholson is now able to produce the frequency of spectral lines from first principles without any data fitting. He applies his hypothesis to protofluorine and predicts fourteen previously observed but unidentified lines in the Sun's corona with an error of less than .4 percent. Turning to nebulium, he predicts ten unexplained nebular lines with an error of less than .01 percent.

Everyone was very impressed, including Bohr. The two men had met in Cambridge in 1911 (before Nicholson had introduced the quantum hypothesis). At the time Bohr wrote, "I also had a discussion with Nicholson; he was extremely kind, but I scarcely agree with him about much." After Nicholson announced the quantization of angular momentum, though, Bohr had reason to change his mind. On a Christmas card to his brother, Harald, in 1912, Bohr wrote, "Although it does not belong on a Christmas card, one of us would like to say that he thinks Nicholson's theory is not incompatible with his own."

Clearly Bohr was struggling with Nicholson's theory, a struggle made manifest in a long letter to Rutherford on January 31, 1913, in which he wrote, "The theory of Nicholson gives apparently results which are in striking disagreement with those I have obtained."* Within a month, though, he had wrapped it up. As unbelievable as it seems, it was evidently only in February that Bohr first encountered the famous Balmer formula. But a glance at the "well-known" formula (as he refers to it in his published paper) was enough to tell him how to combine it with his earlier hypothesis about the stability of the Rutherford atom and Nicholson's proposal to come up with the model all physics students know today: Electrons in the hydrogen atom travel only on stable orbits in which their angular momentum occurs in integer multiples of Planck's constant (divided by 2π). When the electrons make a quantum leap between two such orbits, light is radiated with an energy given by Planck's theory, and the allowed frequencies are exactly those given by the Balmer formula.

Nicholson's fate was hard all around. Despondent about the lack of credit he had received for his contribution, he increasingly took to the wine cellars of Balliol College, Oxford, and ended up hospitalized and forgotten for the last twenty-five years of his life. Recognition permanently passed Nicholson by. One thing that interests me about this Panopticon exhibit is not only how textbook writers have totally misrepresented the history, but how dismissive physicists have been of a "runner-up." There can't be any question that Nicholson's theory was wrong, as became apparent soon enough, but there is also no question that Bohr relied on his critical suggestion. In 1914, before the victor had been declared, the physicist James Jeans wrote that whatever the final outcome, Nicholson's theory had

---

* Heilbron and Kuhn, in "The Genesis of the Bohr Atom" (*Historical Studies in the Physical Sciences* 1, 211–291 [1969]), give this passage in brackets as: "[Nicholson's theory closely resembles mine, yet our results appear at first to be irreconcilable.]"

"probably already succeeded in paving the way for the ultimate explana-tion of the line spectrum." Jeans, though, was far more generous than subsequent commentators. Leon Rosenfeld, one of Bohr's closest collabo-rators, wrote:

> Bohr did not learn of Nicholson's investigations, as we shall see, before the end of 1912, when he had already given his own ideas of atom structure their fully devoloped form. If we have described Nicholson's work in such detail . . . it is in no way to present [it as a precursor] to Bohr's ideas, but rather as evidence of how far able physicists, following the general trend of thought of the time, could carry speculation on these lines. By contrast, the thoroughness of Bohr's single-handed attack on the problem and the depth of his conception will appear still more impressive.

Strangely, while dismissing Nicholson's results as "numerology," Rosen-feld concedes that Bohr had no idea of how to attack atomic spectra in the middle of 1912. Abraham Pais, an eminent physicist and author of a well-known biography of Bohr, is less miserly. He acknowledges Nichol-son's contribution but bends over backward to argue that it was with the stability assumption that Bohr introduced "one of the most audacious postulates ever seen in physics." Historians of science have on the whole been more inclined toward Nicholson. Jagdish Mehra and Helmut Rechen-berg, for instance, write, "All Bohr had to do was take over Nicholson's assumption concerning the discreteness of angular momentum and apply it systematically to determine all possible states of the hydrogen atom." The relative weight one assigns to the ingredients of the physical theory is often a matter of judgment. Although Bohr arrived at the stability postu-late first, it would have been possible to go the other way: Once you accept that orbits must have fixed values of angular momentum, electrons can't go anywhere and stability pretty much follows. In any case, most physicists *believe* the quantization postulate to be the crucial one, in which case Nicholson deserves a kinder reward than oblivion.

What is I think more clear-cut is that there are no villians in this tale, just heroes. Bohr knew of Nicholson's work; he used it to good advantage and acknowledged it. Priority disputes, as often as not, arise after the fact, engendered by supporters of the various actors. However, the type of atti-tude expressed by Rosenfeld in particular not only has by constant repeti-tion engendered the myth of theories sprung full-blown from the head of Zeus but also implies that if you aren't Einstein or Bohr, you might as well abandon scientific pursuits before you start. Science can live without such an attitude, but it can't live without the Nicholsons.

II

The old quantum theory, of which the Bohr model is part, is found in the Panopticon's treasure trove of jury-rigged theories, in this case consisting of a curious blend of quantum concepts grafted onto a Newtonian picture of the atom. But separated from the old quantum theory by a short distance in space and time is the Panopticon's vast expanse of quantum mechanics itself, "new quantum theory," created by Werner Heisenberg and Erwin Schrödinger in the years 1925 to 1927. The trouble with quantum mechanics as they conceived it was that it was incompatible with Einstein's special theory of relativity. A minor shortcoming. Within a year, though, Paul Adrien Maurice Dirac succeeded in wedding the two theories to create a relativistic quantum theory of electrons. His two papers on this subject, published in 1928, are considered among the greatest achievements of twentieth-century physics, and if any physicist receives as much adoration as Einstein among acolytes for pure brainpower, it is Dirac. Dirac stories abound in physics, mostly revolving around his legendary "taciturnity." Probably the most famous concerns a conference a few decades back that took place, if memory serves, in Miami, where Dirac worked in later life. After Dirac gave a lecture, a member of the audience rose and said that he didn't understand one of Dirac's points. He waited for a response. Nothing. More waiting. Silence. After a full minute, the moderator asked whether Dirac intended to answer the fellow's question. Dirac replied: "I didn't hear a question. He only said there was something he didn't understand."

An even more widespread story can be found in any textbook on twentieth-century physics. There you will be told that on the basis of his relativistic quantum theory Dirac predicted the existence of antimatter, specifically the first antiparticle, that is, the antielectron, also known as the positron. So many different books explain this in such hyperrealistic detail that it seems churlish to make an example of *Modern Physics,* but one can hardly think of a more hilarious case of "he who hath, gets," or a more direct proof that nobody bothers to read Dirac's papers anymore. The story goes like this. The equation at the center of Dirac's theory not only described the electron but had a mirror solution that seemed to describe particles identical to electrons but with negative energy. Dirac termed these negative-energy particles "holes," and in 1932 they were accidentally discovered by Carl Anderson, who, not being terribly familiar with Dirac's work, adopted the term *positron* and won the Nobel prize.

What is so funny about all this is that Dirac himself, being by all accounts an intellectually honest individual, admits he got it completely wrong. In the first place, there is no mention of antiparticles in the two famous 1928 papers; those are devoted entirely to the fundamentals of the theory. The "prediction" comes two years later, in a 1930 paper Dirac titled "A Theory of Electrons and Protons." Protons? You must understand that in 1929, when Dirac wrote the paper, only two subatomic particles were known, the electron and the proton. When Dirac found his negative-energy solution, he realized that it could be described by a particle with a *positive* charge. The only particles around with positive charges were protons, and in his paper he quite clearly states "the holes . . . are the protons." He never says anything about antielectrons or antimatter.

But don't take my word for it. Dirac, in his Oppenheimer Memorial Prize acceptance speech, tells the whole story with the clarity for which he was famous:

> One can see at once that such a hole will appear as a particle. It will be a particle with a positive charge and a positive mass. From the begining when I had this idea, it seemed to me that there would be symmetry between the holes and the electrons and therefore the holes must have the same mass as the electrons. How could one then interpret the holes? They would be particles of positive charge. The only particles of positive charge known at that time were protons. For decades physicists had been building up their theory of matter entirely in terms of electrons and protons. They were quite satisfied to have just these two basic particles. The electrons carry the negative charge, the protons carry the positive charge. That was all that was needed. Rutherford had put forward some tentative ideas that there might be a third particle—a neutron. That was just a speculation which people talked about occasionally, but nobody took it really very seriously.
>
> On this basis, that the only particles in nature are electrons and protons, it seemed to me that the holes would have to be the protons. And that was a great worry because the protons have a very different mass form the electrons. They are very much heavier. How could one explain this difference in mass?
>
> I searched about for some time for some cause that would explain it. I hoped that perhaps the Coulomb force* between the electrons might lead to some relationship between all the electrons in the negative energy states which could lead to a difference in mass, though I could not see how it could come about. But still, I thought there might be something in the basic idea and so I published it as a theory of electrons and protons, and left it quite unexplained how the protons could have such a different mass from the electrons.

---

* This is just the electrical force between electrons.

This idea was seized on by Herman Weyl. He said boldly that the holes had to have the same mass as the electrons. Now Weyl was a mathematician.* He was not a physicist at all. He was just concerned with the mathematical consequences of an idea, working out what can be deduced from the various symmetries. And this mathematical approach led directly to the conclusion that the holes would have to have the same mass as the electrons. Weyl just published a blunt statement that the holes must have the same mass as the electrons and did not make any comments on the physical implications of this assertion. Perhaps he did not really care what the physical implications were. He was just concerned with achieving consistent mathematics.

Dirac goes on to explain that Robert Oppenheimer himself voted for Weyl's proposal, that the holes could not have anything to do with protons, and hypothesized that for some unknown reason antiparticles were not very prevalent in nature. That turned out to be true; for reasons not entirely clear to this day, antiparticles are not normally found in our universe. But they are produced in particle accelerators or occasionally in cosmic ray showers, which is where Carl Anderson detected them in 1932. The rest, as they say, is history, or perhaps nonhistory.

* Herman Weyl (1885–1955) is famous for introducing mathematical techniques into quantum mechanics, in particular techniques that describe the symmetries of a system.

10 / A Third Quantum Tale: Southpaw Electrons and Discounted Luncheons

> As I thought about it, as I beheld it in my mind's eye, the goddamn thing was sparkling, it was shining brightly! As I looked at it, I felt it was the first time, and the only time, in my scientific career that I knew a law of nature that no one else knew. No, it wasn't as beautiful a law as Dirac's or Maxwell's, but my equation . . . was a bit like that. It was the first time I had discovered a new law, rather than a more efficient method of calculating from someone else's theory. . . . This discovery was completely new, although, of course, I learned later that others had thought of it about the same time or a little bit before, but that did not make any difference. At the time I was doing it, I felt the thrill of a new discovery! It wasn't as wondrous as Maxwell's equations, but it was good and I was satisfied to sign it. I thought, "Now I have completed myself!"

This is Richard Feynman's celebrated exultation upon discovering a new law of nature, a law that governed the behavior of electrons at the most fundamental level. It is a fine description of the joy of a new discovery, the profound satisfaction that comes at the moment of enlightenment, and from Feynman's words you might well gather that he considered this moment to be one of the supreme moments of his life, if not *the* supreme moment. Yet one of those "others" that thought of it "a little bit before" tells a very different version of the story. It's an instructive lesson with which to end today's excursion to the Domain of Physics and Astronomy of the Panopticon, for it brings us into the mid-twentieth century, well into the range of memory of many people still on their feet, and it is a living illustration of how personal foibles and interactions get ironed into history.

The "others" that Feynman refers to is E. C. George Sudarshan, now at the University of Texas, Austin. I should acknowledge from the outset that I am well acquainted with George, having been a student of his in graduate school and having since coauthored three or four scientific papers, and a popular book with him. Not nearly as famous to the general public as Feynman, George's name occasionally leaks out beyond the walls of academia in connection with tachyons, those hypothetical particles that travel faster than the speed of light, which he invented. Some years ago, at my instigation, *Discover* magazine ran a cover story on George and tachyons that unfortunately left the impression he was something of a nutcase. This is far from the truth. George is probably India's most distinguished physicist and has coauthored over four hundred scientific papers and several textbooks. Born Ennackel Chandy George, he later took the surname Sudarshan, which means approximately "far vision." This was not exactly the act of a shrinking violet, but the plain fact is that Sudarshan is an extremely brilliant man. It is difficult to imagine that many people walk the earth with such a command of physics, and anyone who has ever met him has been awed by the speed of his mind. Now that he is seventy-two he has slowed down to the point where he is only about twice as fast as the rest of us. I count as moments of supreme satisfaction when I have caught him at a mistake, and it has not been often. His sense of humor is pretty good too, and he'd make a great wandering storyteller. George was a pioneer in the field of quantum optics, and for his numerous contributions to theoretical physics he has been well honored in India: At least one institute is named after him, and he is a recipient of the Order of the Lotus, one of India's highest civilian honors. George may also hold the record for the number of times a person has been nominated for the Nobel prize without receiving it. Which gets us back to the subject of this tale.

Before 1957 a standard question on a graduate qualifying exam was "Given only a radio, describe how you would explain to an alien on another planet the distinction between left and right." The correct response was, "It's impossible," and if you think about it for a minute the answer is obvious: All our conventional designations for left and right depend on the direction you're facing. Hence the driving instructor's admonition, "The other right!" when you've turned the wrong way while backing up. If you can't explain to an alien what direction you're pointed, you can't explain to it the difference between left and right.

Back in the early days of rock and roll, nature did not seem to distinguish left and right either—at least the fundamental forces of physics didn't. Gravity makes no distinction between left and right; neither does electromagnetism; nor does the strong nuclear force, the force responsible for holding the nucleus of the atom together. In technical terms we say these forces are "parity-conserving." But in 1957 everything changed. Left out of this list is the "weak" nuclear force. As its name implies, the weak force is *weak*, about a hundred billion times weaker than the strong nuclear force, and it is the force that governs certain types of radioactive decay, for example, the decay of a neutron into a proton plus an electron plus an antineutrino. In the mid-twentieth century the weak interaction was the least understood force in nature, and in 1956 Tsung-Dao Lee and Chen Ning Yang pointed out that no evidence actually existed for parity conservation in the weak force—physicists had merely taken it for granted—and the two men went on to propose several possible experiments to check it.

Nobody believed them. Feynman himself placed fifty-to-one odds against. Yet late the same year a number of research groups, most famously that of Chien-Shiung Wu and her colleagues at the National Bureau of Standards and Columbia University, verified Lee and Yang's conjecture. In a spectacular experiment Wu showed that during the decay of radioactive cobalt, nature did distinguish between left and right. When cobalt decays, electrons are emitted. Now, it happens that electrons can be left-handed or right-handed. Electrons spin like tops; right-handed electrons are those that advance like ordinary right-handed screws, while left-handed electrons advance like screws with left-handed threads. If nature didn't prefer one over the other, you'd expect the same number of each to be emitted in the cobalt experiment, but as it turned out, far more left-handed electrons were emitted than right-handed electrons. The results caused a sensation. "I cannot believe," sputtered a flabbergasted Wolfgang Pauli, "that God is a weak left-hander."

Apart from providing a method to explain the distinction between left and right to an alien—you radio it with directions for setting up the cobalt experiment, and the majority of electrons emitted will be designated left-handed—Wu's experiment also made Lee and Yang world-famous. They shared the 1957 Nobel prize, and since then their names have been inextricably linked, despite the fact that they haven't spoken to each other for decades.

But the weak force shows up in more than the radioactive decay of the neutron. It participates in numerous subatomic processes, which in the 1950s were very puzzling and, according to experimental evidence, contradictory. The question was how to explain the rules for the weak interaction, in other words, to devise a theory of the weak nuclear force. It was perhaps the outstanding challenge in particle physics of the day.

At the time, George Sudarshan was a twenty-six-year-old graduate student at the University of Rochester, having come from the Tata Institute of Fundamental Research in Bombay at the invitation of Robert Marshak, a well-known particle physicist. After Wu announced the parity-violating results in December 1956, Marshak put Sudarshan on the problem. Within a few months Sudarshan had come up with an explanation of the weak force, which for arcane reasons they dubbed the "V–A" theory (V minus A).* The proposal was a rather bold one, for it required "murdering" several experimental results, including one of Wu's, who was considered the most fastidious experimentalist of the time. Sudarshan had hoped to announce the results at the Seventh Rochester Conference in April 1957, but so many people were in attendance that the organizers instituted a rule barring graduate students from speaking. Marshak felt constrained as well. He was chairman and scheduled to give a major talk on another topic. They asked a third physicist to present the results, but he didn't. So Sudarshan's theory, and with it a new law of nature, went unannounced.

The results were presented at another conference in September of that year, the Padua-Venice conference. However, by then . . .

As Feynman tells the tale, by the Rochester Conference he was also fired up about the weak interaction problem and gave a short presentation with some of his conjectures. This he did indeed. But they were just conjectures, conjured up the previous day. Why Sudarshan didn't tell him he had the whole thing wrapped up, I don't know (probably he was too shy), but Feynman went away for the summer to Brazil. On his return to Cal Tech he was told by colleagues that the experimental situation was "such a mess that Murray [Gell-Mann] even thinks it might be V–A." Suddenly Feynman jumped out of his seat, exclaiming, "I understand everything!" He went home and began to calculate. By the next day he had his new law of nature. At that point Murray Gell-Mann, who had been away on vacation in northern California, came back and told Feynman that he was also going to write a paper on the V–A idea. The two of them decided to collaborate.

What this version of the story leaves out is that in mid-June, Marshak and Sudarshan, who had come to California for the summer, invited Gell-Mann to lunch. Nearly twenty years later Gell-Mann claimed that at this lunch he had mentioned the V–A theory as a possible "last stand." However, the other four people at the "summit meeting" had very different recollections. Rather, Sudarshan and Marshak gave Gell-Mann a complete rundown of their theory.

---

* V–A stands for "vector minus axial," which are technical terms designating the type of fields involved in the interactions.

Marshak presented their paper at the Padua-Venice Conference and sent out preprints from Rochester. On that very day the *Physical Review* received an article by Feynman and Gell-Mann.

The rest can be imagined. Because Feynman and Gell-Mann's paper appeared first and in *Physical Review,* for many years they received virtually all the credit for the original theory of the weak nuclear force. The Sudarshan and Marshak paper did appear in the Padua-Venice conference proceedings, which nobody read. The great Hans Bethe was fairly harsh on Marshak, who should have known better about where to publish a paper: "Unfortunately, Marshak and Sudarshan did not write it up as a paper for a normal journal . . . nobody ever bothers to read the proceedings of a conference afterwards, and I think this was just bad luck."

The bad luck did not end. Marshak had sent Robert Oppenheimer their paper in September, and Oppenheimer had replied that he preferred it to the Feynman–Gell-Mann approach, but nevertheless in public referred only to the "Feynman–Gell-Mann theory." A decade later, shortly before his death, Oppenheimer would apologize. "It is a beautiful paper," he wrote Marshak, "and, for whatever good it is, even at this late date I read it with excitement and great pleasure."

Oppenheimer's belated apology was perhaps not much good. This was a major discovery; Feynman considered it the pinnacle of his career, and it would prove to be Sudarshan's best work as well. The V–A theory would eventually be subsumed by the model of Steven Weinberg and Abdus Salam, which unified the weak force with electromagnetism into an "electroweak" interaction, and by the end of the twentieth century, everyone associated with the weak force would win a Nobel prize except for Sudarshan. For years George remained extremely bitter about it all. He once told me that the only way he could gain his freedom was to win the prize, to which I replied, "That's like waiting for lightning to strike." I also happened to see him the morning Steve Weinberg's Nobel prize was announced. George wasn't too happy about it, and isn't to this day. "I still think it was very unfair. To give the second-story man the prize before the first story is given is very unfair." Sudarshan himself expended a good deal of effort lobbying for the prize and even now remains extremely sensitive to any question of priority.* In graduate school we were warned around the department to refer to the Sudarshan–Marshak theory, rather than the Feynman–Gell-Mann theory.

* See note to chapter 7, p. 73.

There can't be any doubt about what happened from Sudarshan's point of view. As late as 1995 we were recording some conversations between us, and the subject turned to the V–A theory and the luncheon with Gell-Mann. George was to the point: "He basically stole it from us. I don't think it is the case that he was waiting to steal something from somebody or other. No, he was concerned about this particular problem. He was minding his business; we invited him to lunch and told him the story from beginning to end. So he talked to Feynman and they wrote a paper."

At that point, George's wife, Bhamathi, jumped in, laughing. "And then Feynman writes in a book that he did it him all by himself. Isn't that interesting?" I pointed out that Feynman did eventually own up to George's priority, and Bhamathi replied, "Feynman does say different things at different times."

Apparently. Whether Feynman knew of Gell-Mann's lunch meeting with Marshak and Sudarshan is, I suppose, the irresolvable gap in his version of events. In reacting to Feynman's account of the moment of discovery, even Gell-Mann says, "It seems an unreasonable conclusion that the whole thing was really so original since Felix Boehm had told Feynman that several of us already had this idea." (Boehm, in fact, was one of the people at the ill-fated luncheon, who had a different recollection than Gell-Mann, and so it's not clear that this statement works entirely in Gell-Mann's favor.)

Decades later Feynman, like Oppenheimer, did try to make amends, as I had mentioned to Bhamati. At one conference he said, "I would like to say where we stand in our theories of weak interactions. We have a conventional theory of weak interactions invented by Marshak and Sudarshan, published by Feynman and Gell-Mann, and completed by Cabibbo." It may seem like a left-handed compliment—after all, the Sudarshan–Marshak paper *was* published, even if Feynman never read it—but in 1985 Feynman wrote to Marshak,

> It was great seeing you and talking. I hope someday we can get this straightened out and give Sudarshan the credit for prority that he justly deserves.
>
> Truly the paper Murray and I wrote was a fully joint paper resulting from many interchanges of ideas between us and it is hopeless to try to disentangle who did what. Nobody should try to determine sources of credit among authors of a joint paper.
>
> These matters all vex me—and I wish I had not caused you and Sudarshan such discomfort. At any opportunity I shall try to set the record straight—but nobody believes me when I am serious. . . .
>
> Best regards, good friend, Dick Feynman.

A skeptic might point out that thirty years after the fact Feynman is still skirting around the main issue—whether he knew of Sudarshan's work at the time he wrote his paper—but this is the best we have, and that's where it rests.

The V–A affair is perhaps good prima facie evidence that the Nobel prize should be abolished—or inflated, as Tullio Regge would argue—and by this time you may well have grown weary of all the clambering after the thing. Outsiders indeed tend to find priority disputes surprising and tedious, but the suprise at least would be diminished if the public would abandon its naive picture of science as a collection of discoveries made by isolated geniuses. Everyone is percolating; a few boil. As for the tedium, well, in the *Bhagavad Gita* it is written that "one has the right to one's labors, but not to the fruits of one's labors," which is the Hindu way of saying "either you do the calculation or you get the credit." It is a hard maxim to live up to, especially in science, where, apart from satisfaction, credit is almost the only thing one can hope for.

My own feeling about the whole thing is that memory is often unreliable and more often convenient. Also true is that people of Feynman and Gell-Mann's caliber have omnivorous appetites and don't pay too much attention to whose toes get stepped on. Others regard Sudarshan himself as not so different in this regard and have sometimes accused him of stealing their ideas. In any case, I will not judge Feynman; he can't defend himself now. But it is a good warning that what you read about cultural icons should be sprinkled liberally with sodium chloride.

With this cautionary tale, we abandon the Domain of Physics and Astronomy. One could, of course, spend a lifetime lost in its ever-convoluting recesses, but our initial excursion has undoubtedly left you exhausted by the abstract, and so we now enter the realm of the concrete, the Domain of Technology. And if you thought physics and astronomy were vicious . . .

II

The Domain of Technology

Like the Physics and Astronomy Domain of the Panopticon, the Technology Domain overflows with priority disputes, error, the mists of legend, and downright lies. The Technology Domain, however, boasts a collection of disputes brought to unsurpassed viciousness by two key factors: fame and money. In pure science, the odds of winning fame or fortune by a brilliant discovery are small, if not infinitesimal. The best you can hope for is the esteem of your colleagues and perhaps, if you live long enough, a medal. Not so with technology. A great invention can change the course of civilization and, incidentally, make you millions. Consequently, the history of technology is filled with patent claims, counterclaims, and lawsuits, many of which have dragged on for decades.

And which go back as far as records exist. Survey, for example, one of the most prominent of the Panopticon's exhibits, the invention of the printing press, which was the most important advance in history, we are told. Although universally attached to the name Gutenberg, this is less a fact than a place saver. About the only thing known with certainty is that by 1450 or so Johann Gutenberg of Strasbourg and his partner Johann Fust had perfected movable type to the point that they could begin to print the Vulgate Bible. Gutenberg's name does not appear on any of the works printed by him. This may be due in part because he had failed to repay a loan to him by Fust, who then brought suit, and in the settlement Gutenberg was required to renounce all claims to the invention. But the origin of the printing press is treacherous all around. Even in current encyclopedias can be found the name Laurens Janszoon Coster of Haarlem, whom the Dutch advance as the true inventor of movable type; however, this claim appears to be totally legendary, invented over a century after the fact. In the northern Italian town of Feltre stands a monument to Pamfilo Castaldi of Milan, who supposedly preceded the others, but there seems to be no more evidence in favor of him than Coster.

Certainty be damned. Paul Needham, librarian of Princeton University's Scheide Library, and Blaise Agüra y Arcas, a 1998 physics graduate, have recently made a digital comparison of the individual letters of the Scheide Library's Gutenberg Bible and other contemporary documents. They conclude that these works were not printed by letters cast in metal molds of the type Gutenberg is supposed to have invented. Rather, they were printed by an earlier technique that involved casting letters in non-reusable sand molds. The two men now believe that movable type was probably invented a few years after Gutenberg's death in 1468, presumably by someone else. Of course, all of this ignores the Koreans, who introduced movable type at least fifty years before the Europeans, and the Chinese, who invented printing.

More clarity but equal confusion surrounds another great invention. Despite what every American schoolchild is taught, neither James Watt, Robert Fulton, John Fitch, nor Simon Newcomen invented the steam engine. Some ten or fifteen years prior to Newcomen, in 1698, Thomas Savary demonstrated his "miner's friend" for King William III and within the year received a patent for it. Little is known about Captain Savary, but he was a prolific inventor, and the "miner's friend" seems to have been the first steam engine in the modern sense of the word—a device that continuously cycled to perform its work. This original "fire-engine" was used to pump water out of mine shafts by means of two large copper tanks that were alternately heated and cooled as steam passed through them. Water in one of the oval vessels was expelled upward as steam entered from below through a pipe. The vessel was cooled, condensing the steam and creating a partial vacuum. As in Torricelli's (Viviani's?) barometer, atmospheric pressure then forced water from the mine below into the tank through a second pipe, and the process was repeated. The use of high-pressure steam made the "miner's friend" extremely difficult to operate for the day.

Newcomen, whose steam engine was also used to pump water, seems to have devised it independently of Savary, but in fact he teamed up with his predecessor because the captain's patent was so broadly written that it blocked the production of almost any conceivable "fire-engine." Unlike the "miner's friend," Newcomen's device used a cylinder and piston based on the 1690 design of Denis Papin. Newcomen and Savary erected the engine at Dudley Castle, Worcestershire, in 1712. It was another fifty years before Watt got into the act and a full century before Fulton sailed his steamboat, the *Clermont,* up the Hudson.

Fulton, being an American, is of course featured in texts much more prominently than the others, even more prominently than his compatriot

John Fitch, whose earlier boat, first tested in 1786, was a financial failure. But in fact James Rumsey of Virginia built a steamboat at about the same time, in 1785. Rumsey's boat employed the jet principle—steam ejected from nozzles propelled the boat forward—but the initial test in early 1786 was also a dismal failure, and presumably for that reason Rumsey is mentioned even less often than Fitch in textbooks—and an order of magnitude less often on the Internet. What goes missing is that Rumsey went on to improve the boiler, and at Shepherdstown on the Potomac he made a considerably more successful trial run. Whether this test took place in 1786 or late 1787 has been the subject of considerable and vicious debate. In any case, a Rumseyan society was soon formed in Philadelphia, whose members—including Ben Franklin—attempted to promote the inventor's activities. Perhaps Rumsey would be remembered as the father of the steamboat if he had not gone off to England to raise funds and died there. Perhaps not. Before Rumsey's departure, a priority dispute between himself and Fitch broke out. Pamphlets flew this way and that in which Rumsey accused Fitch of perfidy and Fitch accused Rumsey of antedating his first trials. Presumably the matter hinges on whether Rumsey's first, failed trial should be counted, and it's not clear the matter has ever been sorted out. Despite George Washington's own testimony that Rumsey had discussed the idea with him prior to Fitch, Fitch was eventually awarded a patent.

Why stop in eighteenth-century America? It's accepted that Marquis Claude Jouffroy d'Abbans (1751–1832) successfully operated a paddle-wheel steamer, the *Pyroscaphe,* on the river Saône in 1783, and other attempts were made even earlier. An interesting tale surrounds frequent claims found in texts and now on the Internet that the first steamboat was actually operated at the improbably early date of 1543 by a captain in the Spanish navy, Blasco de Garay. Details of Garay's mechanism do not survive, but a description of the trials was published in 1825 by the superintendent of the Spanish Royal Archives based on manuscripts contained therein. According to this account, Garay proposed to the Holy Roman Emperor Charles V a ship that would propel itself "even in calm weather" without oars or sails. "Garay would not explain the particulars of his discovery: it was evident however during the experiment that it consisted in a large copper of boiling water, and in moving wheels attached to either side of the ship. The experiment was tried on a ship of two hundred tons called the *Trinity,* which came from Colibre to discharge a cargo of corn at Barcelona."

The Royal (now National) Archives at Simancas have been a mainstay for historians for centuries, and the published account is sufficiently detailed that one tends to believe it. However, the matter was evidently cleared up as early as 1858 when historian John MacGregor visited Spain and, not trusting the superintendent's report, actually read Garay's original letters at Simancas. MacGregor reports that the letters "gave particulars of experiments at Malaga and Barcelona, with large vessels propelled by paddle-wheels turned by twenty-five *men*" (italics mine). As for the steam engine, MacGregor tells us, "After careful and minute investigations at Simancas, Madrid and Barcelona, I cannot find one particle of reliable evidence for this assertion." According to the historian H. Philip Spratt, the cauldron on deck, which gave off steam, seems to have been used for defensive purposes. Spanish authorities themselves discredited the claims for Blasco de Garay's invention the better part of a century ago, but the legend sails on and . . .

Fulton, by the way, was well aware of his many predecessors.

From this brief account, three lessons stand out. One is that the criterion for invention is even less clear-cut than that for discovery in pure science. People often cite patents as proof of invention, but this is a legal, not scientific, criterion whose conditions vary from country to country. As will become ever clearer in the Technology Domain of the Panopticon, scientists and inventors often patent ideas without building actual devices, often idealistically demonstrate "proof of concept" by documented experiments but without bothering to patent, and perhaps just as often retrospectively report such experiments from memory without documentation but sometimes with credible eyewitness accounts. Who is the inventor? Whatever the criterion, one should stick to apples or oranges. All too often a partisan will dismiss a claim on grounds such as "Popov demonstrated radio before Marconi patented it." In that case, one should ask when Marconi first performed a successful experiment.

The second lesson is that patent laws at one time may have been good for inventors and better for lawyers, but they are not good for history. The very act of awarding a patent singles out a particular individual as "the" inventor and excludes everyone else, which exerts a tremendous distorting force on our view of the actual course of events. As in pure science, it is rare that one person is far ahead of anyone else, and if he is, he is usually ignored.

The third obvious lesson from the steamboat tale is that failure is not a terribly popular exhibit in the American Panopticon. For that reason, when one points out that Edison did not really invent the lightbulb, patriots reply, once they get over the shock, that it must have been the best lightbulb, which is what counts. No one would deny that Edison's bulb was the best, but by that argument, Stephenson should not be considered the inventor of the locomotive, which traveled only about two hundred yards before it broke down the first time.

We routinely mistake a superior design for the original concept. This is clear in the case of Eli Whitney, who has been immortalized as both the inventor of the cotton gin and the inventor of interchangeable parts. He did neither. Although Whitney's cotton gin ("engine") was probably the most important technological development in early America, and thousands of scholars have written studies about its impact on the course of slavery and all history, it was certainly not the first; it was the best. Previous gins existed in the Orient and India, and their basic principles had been imported to Louisiana by 1725. These early gins, simple things that resembled clothes wringers, performed well enough on Asian cotton but were not effective in separating the seeds from the fiber in American cotton. Whitney, the story goes, within ten days after arriving at the Georgia plantation of Catherine Lidfield Greene, solved the problem by creating a gin in which toothed rollers pulled the fiber through slots, separating it from the seeds.

The best, then, if not the first, but in the last few years claims have arisen that Whitney does not deserve credit for even that much. Seale Ballenger in *Hell's Belles* writes that it was Greene, the widow of the revolutionary war hero Nathanael Greene, who "not only thought of the idea and told Eli about it, but let him live in her house, supported him while he perfected it, and helped make modifications to it for him." Nevertheless, an ungrateful Whitney patented the gin under his own name with no acknowledgment of her help. Dozens of such claims—some contending that Whitney stole the invention outright from Greene—have also been posted on the Internet; however, neither Ballenger nor any of the rest of them provides a shred of documentary evidence.

More substantiated is the claim that Whitney did not invent interchangeable parts. The archaeologist David Starbuck, who participated in the excavation of Whitney's gun factory in New Haven, reports that Whitney never claimed to have invented interchangeable parts (he called them "uniform") in the first place; those assertions appeared only after his death. Whitney never disclosed his manufacturing methods, and despite the fact that he was supposedly using uniform parts, it took him ten years to fill a government order for ten thousand muskets. Working at the Smithsonian, Starbuck's colleague Edwin Battison disassembled some Whitney muskets and found they had been individually filed and numbered. Starbuck's excavation revealed a complete absence of the machinery Whitney was supposed to have invented (drawings of which nevertheless appear in high-school texts) and no parts whatsoever. Starbuck concludes that Whitney was ultra-secretive about his gun inventions precisely because there weren't any. What remains to be understood is how the whole rigamarole got started.

By this point in the Panopticon tour it is clear that nationalism is a fundamental concept, both present and past. Yet the depth to which it has penetrated and the shock of being confronted by alternative histories is never brought home more forcefully than by a trip to Russia. When visiting the (now closed) Soviet branch of the Panopticon, we hear, a certain tourist was being guided through the Hall of Inventors. The guide pointed to a large portrait on the wall and explained proudly to the visitor, "This is Popov, who invented the radio before Marconi." Next to it hung an equally large portrait, and the guide announced, "This is Mozhaisky, who invented the airplane before the Wright brothers." At a third portrait she said, "This is Lodygin, who invented the lightbulb before Edison." And so it went, down the entire gallery, until it seemed that Russian inventors had created all modern civilization. As the pair reached the end of the hall, a portrait larger than any of the previous loomed before them, but now the tour guide fell conspicuously silent. "Well," said the visitor, "aren't you going to tell me who this fellow is?" "That," the guide answered matter-of-factly, "is Ivanov, who invented the rest of those guys."

During the Cold War it was standard procedure for the Soviets to claim priority in every sphere, and standard procedure for Westerners to reject the claims as propaganda. But as they say, just because you're paranoid doesn't mean they aren't out to get you. The fact is Russia did produce outstanding inventors who remain relatively unknown in the West. And although to this day Russians tend to overstate their cases for priority, the Cold War is long gone and the time has come for a more objective evaluation of their claims. As we will see during the approaching excursion, one can hardly deny that some of these individuals made significant contributions toward the telegraph, lightbulb, radio, and television. If their impact on the overall development of these fields was less than the more familiar names, it is largely due to the fact that Russian pioneers worked in an isolated country, often in isolation themselves, and lacked the financial backing of their Western counterparts.

On the other hand, outsiders, especially if they emerge from the shrouded mists of exotic lands, have an undeniable romantic appeal and develop cult followings. Among modern visionaries, Buckminster Fuller is the most famous of these. Anyone will tell you he invented the geodesic dome (why this is considered such an important invention has never been clear to me), but it is a matter of record that such a dome was designed and built by the Zeiss optical company in Jena, Germany, thirty years earlier, and a patent exists to prove it.* (As far as I know the dome still exists to prove it as well.) Of course, the visionary Fuller is a more suitable icon than Walter Bauersfeld, the chief engineer at Zeiss, who merely needed to design a planetarium dome; Bauersfeld nonetheless seems to have been a fascinating and cultured individual in his own right.

Nikola Tesla has always commanded a legion of followers, and recently Hedy Lamarr has joined the ranks. We will meet both of them during today's contemplation at the Panopticon, but let us begin with a true American legend, Samuel Morse and the telegraph.

---

* See my *Science à la Mode* (Princeton: Princeton University Press, 1989) for a full account. The patent as well as photos of the original dome can be found on my Web site (wwwrel.ph.utexas.edu/~tonyr). Walter Bauerfeld's daugher tells me that she brought Bauersfeld's invention to Fuller's attention after a lecture and that he was so shocked, she is convinced he knew nothing about it.

11 / What Hath God Wrought? Shadows of Forgotten Ancestors, Samuel Morse, and the Telegraph

There is something nicely American—perhaps uniquely American—about a tale in which a gifted artist who is at the same time a religious fundamentalist, who displays as much paranoia as he does a lack of knowledge of electricity and magnetism, who ungraciously denies the contributions of his predecessors, and who at the end of the day is immortalized as the inventor of the telegraph. The moral we accept with humility: Time is a harsh mistress, but more so, all sins are forgiven in light of a success.

Let us be clear from the outset: Samuel Morse was a fascinating character. Someone should write a play about him, if not an opera. Let us also be clear: He did not invent the telegraph. The basic ideas that led to the first telegraphs had been known to the ancient Greeks and Romans, in particular that rubbing amber with fur produced a static electric charge and that iron needles stroked with lodestone would attract each other (chapter 3).

The primordial telegraphs took over these observations directly. In his monumental history of the subject, John Fahie details at least a dozen, if not two, telegraphs based entirely on the properties of static electricity. The earliest such device was described in a letter to the *Scot's Magazine* in 1753. The author of the letter, C.M., has never been identified beyond doubt, but whoever wrote it describes a remarkably complete system. C.M. claims to have strung a series of wires "equal in number to the letters of the alphabet" between a transmitting and a receiving station. At the receiving end, he tells us, small balls are attached to the ends of the wires and, lying "about a sixth or an eighth of an inch" below each ball, is a letter of the alphabet scribbled on a bit of paper. C.M. then gets his "electrical machine" a-goin', charges a glass rod with it, then strikes the *s* wire, the *i* wire, and the *r* wire with the glass. At the other end, the static charge on the *s* ball lifts the letter *s* beneath it, and so on, until C.M.'s interlocutor has received the word *sir*. C.M. also recommends the use of small bells, which ring at different pitches when struck by an electric spark, if one finds the bits of papers blowing around tiresome. The intrepid inventor seems to have performed the experiment over a distance of thirty or forty yards.

Most of the proposals for static-electric telegraphs were along those lines. One was described in a Latin poem of 1767; by 1787 there is an eyewitness account that one was actually built by an M. Lomond in Paris. Lomond employed a single wire, which means he must have devised a code for the letters. This in itself was not so unusual; semaphore signals are nothing more or less than a telegraph code, and at least one of the earlier telegraph proposals explicitly mentions an encoding system. Do-it-yourselfers continued to tinker with static-electric telegraphs until well into the 1830s—almost to the day when Morse appeared on the scene. They understood the need to insulate wires and proposed many modern ways of doing so, but none of the experimenters seems to have succeeded in devising a telegraph capable of transmitting signals more than five hundred or so feet.

One might guess that a real electric current, the kind produced by batteries, would be more suitable for transmitting a telegraph message. Indubitably. One immortal day in 1786, the story goes, the anatomist Luigi Galvani and his nephew Camillo had prepared some frogs for experiment and had hung them by iron hooks from the top of an iron balcony outside the laboratory. Galvani noticed that when a frog was blown against the balcony, its legs would contract. Galvani devised elaborate theories to explain this puzzling phenomenon, which gave rise to the term *galvanism*— animal electricity. Today we know that the chemicals in the frogs' legs, when put in contact with certain metals, behave like a battery. It was precisely in trying to establish the exact mechanism behind galvanism that Alessandro Volta developed his own battery in the late 1790s, unveiling it in the year 1800.

And it was Galvani's experiments that led to the first transmitter using an electric current—the frog telegraph. Learning of Galvani's experiments in 1800, Don Francisco Salvá reported to the Academy of Sciences of Barcelona that he had managed to transmit an electric current over a three-hundred-meter wire and succeeded in causing a frog's legs to contract. He expressed the conviction that such a telegraph could be used for communication over far greater distances. The source of the current? Yes, for a battery Salvá employed the "electricity produced by a great number of frogs"! A few years later Salvá built a telegraph utilizing a voltaic pile. On the receiving end, instead of frogs' legs, he placed electrodes in jars filled with water. When the wire designating a particular letter was con-.

nected to the battery, bubbles appeared in the corresponding jar as the water was separated into hydrogen and oxygen. Thus the birth of the bubble telegraph.

S. T. von Sömmerring is frequently credited with the first electrical telegraph, but his device debuted only in 1810 and is a more elaborate version of Salvá's idea, with a large receiving tank of water and thirty-five wires, for all the letters in the German alphabet and ten numerals besides. All the same, Sömmerring's telegraph, which he had set up at his house in Munich, attracted throngs of statesmen, princes, and philosophers, and operated over the "greatest distances." Over the next few years the range of electrical telegraphs was extended to a reported seven miles, and by 1813 the device had attracted enough attention to be the subject of playful verse in *The Satirist:*

> Electrical telegraphs all must deplore,
>   Their service would merely be mocking;
> Unfit to afford us intelligence more
>   Than as would really be *shocking!*

A pivotal event in this tale was Oersted's monumental discovery of 1819–1820 (or Romagnosi's, if you believe it; see chapter 3) that a current-carrying wire would deflect a compass needle, thus demonstrating that electricity and magnetism are intimately connected. Scientists comprehended the implications immediately, and almost as soon as Oersted's discovery became public, the French mathematician Laplace proposed its employment for telegraphic purposes. André-Marie Ampère himself, in a paper read before the French Academy of Sciences, then took over the suggestion and sketched out an electromagnetic telegraph. Depressing a key on an alphabet keyboard would send a current from a battery down the corresponding wire; at the far end, where the wire began its return journey, a compass needle would deflect, indicating which letter had been chosen.

By 1830 William Ritchie at the Royal Institution had actually constructed a telegraph along these lines, improving upon Ampère's plan by adopting a suggestion of Johann Schweigger to wrap the compass needles within coils of wire (which increased the deflection). Ritchie's device, which was exhibited in London, evidently operated over a few hundred feet, but the inventor argued that with larger batteries such devices could transmit signals over four hundred miles. And there's the rub. Other natural philosophers had shown that the current diminished so much over a few hundred feet that they felt certain a practical telegraph was impossible.

*   *   *

The plots begin to converge. If you have stopped by the Joseph Henry exhibit at the Panopticon (see chapter 3), you will know that during these years he was experimenting with electromagnetism in Albany. By 1831 he had set up a demonstration telegraph in his lecture hall, looping wire around the room until students said its length was a mile and a half. A current sent through the wire would trigger an electromagnet, which would in turn cause a suspended bar magnet to rotate and strike a bell. This, as far as we know, was the first demonstration of the possibility of a long-distance telegraph. Later, while a professor at Princeton, Henry would set up a telegraph between his laboratory and his home, which seems to have been one of the first not to require a return wire to complete the circuit—Henry used the earth itself.

At about the same time Henry developed a device today known as the relay, which would lie at the heart of all future telegraphy. A relay is basically a switch operated by an electromagnet, a switch that opens one circuit while simultaneously closing another.* By 1835 he had developed the relay for use in "distance circuits," the idea being that over a long wire a tiny current would trigger a small electromagnet, which in turn would activate a larger magnet, energized by its own battery. Henry used relays to cause his heavy-duty magnets to drop weights. But the same idea could be used to operate a distant telegraph receiver. The professor was quite aware of the commercial possibilities—he even mentioned them in an 1831 article—but chose to devote himself to fundamental research. That error would prove fatal.

Meanwhile in St. Petersburg: One of the myriad visitors who saw Sömmerring's telegraph in action was Count Pavel Lvovich Schilling (1786–1837), then a translator in the Russian embassy in Munich. He and Sömmerring became fast friends, and so impressed was Schilling with the German's invention that he decided to devote his spare moments to the science of electricity. During the Napoleonic Wars, Schilling, who served in the dragoons, contrived what appears to have been the first underwater cables, insulated by india rubber and varnish. With these he detonated mines across the Seine. After the wars Schilling organized the first lithography in Russia as well as expeditions to the Chinese border, and by 1828 he had begun experimenting with a single-needle telegraph.

---

* The electromagnet, when activated by a current passing through it, might pull down one end of a little iron seesaw, which is connected to a battery terminal. The connection to the battery will be broken. At the same instant, the other end of the seesaw, being raised, might connect to the terminals of another battery, putting a second circuit into action.

In 1832 he unveiled a telegraph system, which can still be seen in the Moscow Polytechnical Museum, consisting of six compass needles suspended above magnetic coils. In order to be able to transmit messages with this system, Schilling devised a code, remarkably similar to Morse's, based on two symbols. Officials at the Moscow Polytechnical Museum insist that this device was the world's first electromagnetic telegraph, but in light of Ritchie and Henry's experiments, such a claim seems difficult to substantiate. Schilling may well have been aware of Henry's work.

Nevertheless, Schilling's system was the most advanced of its day, and attracted much attention, including that of Tsar Nicholas I. By 1836 the inventor had tested a line of some nine kilometers running around the Admiralty in Petersburg, and the following year an imperial decree authorized him to connect the tsar's residence in Peterhof to Kronstadt Island. Unfortunately, Schilling died just as the project was about to be implemented, and Russia narrowly missed the opportunity to be the first country with a truly practical telegraph line, as well as a submarine cable. Ironically, Russia became one of the last countries to accept telegraphy, because Nicholas I, so anxious to have a military system, regarded a civilian one as an instrument of subversion.

Schilling's work had a direct impact on that of his English colleagues, in particular on William Cooke and Charles Wheatstone. In 1836 Cooke saw one of Schilling's instruments on display in Heidelberg and upon his return to England was commissioned to install a telegraph along the Liverpool-Manchester railway. The technical difficulties Cooke encountered, however, forced him to team up with Wheatstone, a professor at King's College, London, who had also been working on the telegraph. By now the major problem was widely recognized: how to transmit a signal over long distances. It appears that neither Cooke nor Wheatstone understood how to solve this problem until a visit by Joseph Henry in 1837, when Henry explained the principles of the relay, as well as the nature of resistance in wires (known today to physics and engineering students as Ohm's law). A few months later, in June 1837, Cooke and Wheatstone received their first patent for a five-needle telegraph, and within a few years it was widely employed along British railways. A spectacular success took place in 1845. A murderer was descried boarding a London-bound train at Slough; officials sent a telegraphic signal ahead to Paddington station, and the scoundrel was apprehended and later hanged.

To this day, Cooke and Wheatstone are regarded in England as the inventors of the telegraph. By a fluke. A rival, one Edward Davy, had devised a vastly superior system, complete with relays, ground returns, code, and a method of recording the signals on paper. When he heard that Cooke and Wheatstone had applied for a patent, Davy quickly filed a caveat, a formal declaration that an invention was about to be unveiled. His competitors attempted to quash Davy's work, but in the ensuing court battle the judge solicited the advice of an authority no less than Michael Faraday himself, who testified that Davy's patent should stand intact. Over the next two years, in the race to get the railways to erect telegraphs, Davy's plans more than once came to the verge of fruition, and it seemed that the matter would go hard with Wheatstone and Cooke. But zounds! In 1839 deplorable circumstance of a personal nature drove Davy forthwith to Australia. The loss, acquaintances asserted, was nothing less than a national calamity. And the field was clear for his rivals.

Except: Davy had noted a year earlier in a letter, "I have had notice of another application for a patent by a person named *Morse*. Messrs. Cooke and Wheatstone have entered into an opposition to this application, and I shall have to do the same."

Samuel Finley Breese Morse was born April 27, 1791, into the household of Jedidiah and Elizabeth Breese Morse of Charlestown, Massachusetts. Jedidiah, one of America's most prominent Calvinist ministers, was the architect of the Second Great Awakening, the evangelical crusade of old New England Calvinists against liberal theology and Jeffersonian democracy. The influence of such a patriarch on his son can be imagined but doesn't need to be: Samuel, like his father, came to believe that he had received a mandate directly from God.

That trait, early on, was nowhere to be found. Evident only was a genuine talent for painting, and so in 1811, after graduating from Yale, Samuel set off to London to enroll in the Royal Academy of Art, where he was received by none other than Benjamin West. Samuel's letters home are observant, affectionate, moderate, and intelligent. His initial impressions of Europe were highly favorable; he notes that "no person is esteemed accomplished or well educated unless he possess almost an enthusiastic love for paintings. To possess a gallery of pictures is the pride of every nobleman." By comparison, in America arts seemed "to lie neglected and only thought to be an employment suited to a lower class of people."

Little has changed. After three years abroad Morse returned to America and over the next decade of struggle became one of the most p rominent painters in the United States. One cannot deny that he was a highly gifted, even great artist. Many of his portraits remind one of Gilbert Stuart; some of his more visionary works, of Thomas Cole. His circle of friends included the leading intellectuals of the day, including James Fenimore Cooper and General Lafayette. Determined to make art an integral part of American life, Morse led an 1826 rebellion against the American Academy of the Fine Arts, which was little more than a club for prominent artists who purchased membership by buying shares, a gang of elitists who refused point blank to offer students instruction and who did not even deign to stage changing exhibits. A band of about thirty revolutionaries, including Morse and Thomas Cole, literally seceded in an action that involved disguised messengers, secret meetings, threats, strikes, and lockouts. The result of the sedition was the establishment of the National Academy of Design, the most significant event of nineteenth-century American art. Morse was elected its first president.

By the 1830s, however, Morse's story darkens. His artistic career went into decline. *The Gallery at the Louvre,* a painting on which he had pinned high hopes, was a public failure, and Congress refused to grant him a long-awaited commission to paint a mural for the Capitol rotunda. In 1837, at the age of forty-six, Morse ceased painting forever. By that time, however, he already had other things on his mind.

One was Irish Catholics. Morse had again set sail for Europe in 1829, spending three years there. In his letters from Italy in 1831, one begins to perceive the painter's anti-Catholic fervor. He writes of the Church, "It is a religion of the imagination; all the arts of the imagination are pressed into its service . . . lent all their charm to enchant the sense and impose on the understanding by substituting for the solemn truths of God's Word, which are addressed to the understanding, the fictions of poetry and the delusions of feeling. The theatre is a daughter of this prolific mother of abominations." Of Europe, his early admiration is now nowhere to be found: "America is the stronghold of the popular principle, Europe of the despotic. These cannot unite; there can be, at present no sympathy."

The incendiary situation in New York when Morse returned there in 1832 seemed to have catalyzed his growing antipathy toward foreigners and Catholics into full-blown paranoia. A rising tide of Irish immigrants had dramatically increased the Catholic population, and during the elections of 1834, three days of religious-ethnic rioting brought the city to the verge of anarchy. Morse "deemed it a duty to warn the Christian community against the . . . dangers arising from Popery" and unleashed a flood of venom against Catholics and the pope in the pages of the *New York Observer,* a newspaper run by his brother. The following year he collected his views into a book, *Foreign Conspiracy Against the Liberties of the United States: The Numbers of Brutus.* It is a remarkable performance—a rambling attack against the "Catholic sect" of America, and bigoted and delusional in every sense of the words:

> We have cause to expect an attack, and that it will be of a kind suited to the character of the contest, the war of opinion. . . . *Austria is now acting in this country.* She has devised a grand scheme. . . . She has her Jesuit missionaries traveling through the land; she has supplied them with money, and has furnished a fountain for a regular supply. She has expended a year ago more than *seventy-four thousand dollars* in furtherance of her design! These are not surmises. They are facts. . . . The great patron of this apparently *religious scheme* is no less a personage than the *Emperor of Austria.*

Prince Metternich, "the arch contriver of plans for stifling liberty," and the St. Leopold Foundation were planning an overthrow of the U.S. government. "Must we wait for a formal declaration of War?" Morse asks after two hundred pages. "The serpent has already commenced his coil about our limbs, and the lethargy of his poison is creeping over us; shall we be more sensible of the torpor when it has fastened upon our vitals?"

Foreign Conspiracy ran through eight printings. He followed up the triumph by promoting one Maria Monk's *Awful Disclosures: The Hotel Dieu, Nunnery in Montreal,* in which she claimed to have witnessed unnatural sexual acts in the convent and seen crypts filled with the corpses of illegitimate children. Monk, with whom Morse was rumored to be romantically involved, turned out to be an escapee from a lunatic asylum. Apparently on the verge of losing his sanity himself, Morse was nominated by the Native American Democratic Association to run on an anti-Catholic and anti-immigration ticket for the 1836 mayoral elections. He was, thankfully, defeated by a landslide.

One will not find this aspect of Morse's personality highlighted in most biographies or at the Library of Congress Web site, yet it is crucial to understand that he viewed the telegraph and himself as instruments of God. Morse had been an occasional tinkerer, had encountered electricity during his student days at Yale, and had in 1827 attended a series of lectures on electricity by James Dana at the New York Athenaeum. That was the extent of Morse's scientific training, but apparently it sufficed. The idea for the telegraph, he maintained, came to him on his return voyage to the United States from Europe in 1832. During a shipboard conversation about then-current researches on electricity in Europe, Morse suddenly exclaimed, "If the presence of electricity can be made visible in any part of a circuit, I see no reason why intelligence may not be transmitted instantaneously by electricity." That, at least, is how he reported it in later years. Morse and his partisans always insisted that he worked out the entire system from beginning to end on that voyage, including the relay. Unfortunately, contrary to his usual practice, Morse failed to keep a diary on that voyage, and his early sketchbook of his ideas mysteriously perished in a fire. Thus the window was flung wide for one of the century's great technological scandals.

Over the next four years, while Morse was still engaged in art and politics, he built a few prototypes of his telegraph in his office at the New York University, where he had been appointed professor. What distinguished Morse's invention from those of his predecessors (except perhaps Davy's) was that he eschewed compass needles from the outset. Rather, the plan was to employ an electromagnet to rapidly lift and drop a metal bar fitted with a stylus onto a moving roll of paper such that messages would be recorded. Nevertheless, Morse soon ran into the same problem that so bedeviled his European forebears. As his colleague Leonard Gale, a professor of chemistry at the university, put it, "Morse's machine was complete in all its parts and operated perfectly through a circuit of some forty feet, but there was not sufficient force to send messages to a distance." That year, 1837, Gale acquainted Morse with Henry's 1831 paper on electromagnets, and soon they were transmitting signals through miles of wire wrapped around Gale's lecture hall. On October 3 Morse filed a caveat with the Patent Office.

Here the chronology becomes befogged, all parties suffering selective amnesia. At about that time Morse and Joseph Henry accidentally met in New York. A decade later Henry would describe Morse as "an unassuming and prepossessing gentleman with very little knowledge of the principles of electricity or magnetism. He made no claims in conversation with me to any scientific discovery beyond this particular machine." Nevertheless, Henry, who was in the habit of dismissing would-be inventors, pronounced Morse's telegraph workable, and a cordial friendship ensued. Henry freely gave Morse advice; Morse returned the favor by lending the professor five miles of wire for his experiments. The following spring Morse was off to Europe to secure patents for his invention, where he succeeded in France and failed in England, due to the interference suit brought by Cooke and Wheatstone. Upon his return in 1839 he visited Henry in Princeton, thus renewing their interaction, which continued a further six years. It does strike one as strange, given that Morse claimed to have invented the relay in 1832, to find Henry having to suggest as late as 1839 that "if the length of wire between stations is great, I think that some other modification will be found necessary to develop a sufficient power at the farther end of the line." Even in 1842 we find Morse sending Henry twelve biscuit cups with the request that Henry might use them to make Morse a battery. Equally strange is that Henry does not appear to have set eyes on the invention until the same year, which forces one to wonder not only as to the state of it but as to the extent of the scientist's participation.

Well. Henry continued to support Morse. Cooke and Wheatstone, in fact, were attempting to obtain patents in the United States, and their efforts (which eventually succeeded) put Morse's plans in dire jeopardy. Henry's greatest service to his compatriot was perhaps his 1842 letter of endorsement, with which Morse canvassed members of Congress to secure funding. The strategy bore fruit; in 1843 Congress awarded Morse $30,000 to construct an experimental telegraph line between Washington and Baltimore.* Yet within a year the project was on the verge of collapse. Morse's original scheme had been for an underground cable, but the insulation of the wires proved so bad that the idea failed miserably; he ordered the line surreptitiously torn up. How to insulate the wires? Morse's engineer, Ezra Cornell, came up with the same solution that had been found in Europe: attach bare wires to telegraph poles with glass insulators. Morse actually rejected the idea until Henry told him it was the right one.

And so on May 24, 1844, before an invited audience at the Supreme Court, the daughter of the commissioner of patents dictated the first public telegraphic message. Forty-four miles away, in Baltimore, the message was received: "What hath God wrought?"

What God hath wrought was the nineteenth-century Internet, an invention that bound the country together as never before. Morse himself understood as early as 1838 that "it is not visionary to suppose that it would not be long ere the whole surface of this country would be channelled for those *nerves* which are to diffuse, with the speed of thought, a knowledge of all that is occurring throughout the land, making, in fact, *one neighborhood* of the whole country." He was not far wrong. Within scant years of the Washington-Baltimore demonstration, the country—the world—was crisscrossed with telegraph lines, and Morse had become an international hero.

---

* One should probably mention that the $30,000 was awarded only after the chairman of the House Commerce Committee, Francis Ormand Jonathan Smith of Maine, was made a full partner in the Morse company.

What God hath wrought was also an unprecedented number of law-suits that plagued Morse for years and which had a far-reaching effect on American patent law. One such was the suit Morse brought against Henry O'Reilly, who had started a rival telegraph company. The animosity was so great that at one point armed conflict threatened to break out between the two opposing gangs. Another suit on patent extension reached the Supreme Court. Morse triumphed in virtually every case.

But at a great price, for what God had also wrought was a tragic, irreparable rift between Morse and Joseph Henry. The falling-out began in 1845 with the publication of a book by Alfred Vail, Morse's young part-ner and financial backer, titled *The American Electro-Magnetic Telegraph*. While providing a remarkably detailed—if dismissive—history of virtually every telegraph system constructed before Morse's, it singularly failed to mention Henry's contributions. Henry may not have patented his discover-ies, but he most certainly expected credit. When news of Henry's displea-sure reached Morse in 1846, the latter wrote what is to all appearances a deeply felt and contrite apology, confessing to "feelings of deep regret and mortification" and pleading ignorance of Vail's offense. Henry seems to have been only half convinced, appending a note to Morse's gesture: "This is a very good letter, and I think Prof. M. intended to do me full justice, but when he came to learn the true state of the case he had not the mag-nanimity to do that which was right." Henry's displeasure only increased at the receipt of a letter from Vail, dictated by Morse, in which Vail claimed that he had failed to mention Henry only because he was unable to ascertain the nature of his discoveries. To be sure, in correspondence between Morse and Vail regarding the book, Henry is never mentioned. Whether this was due to oversight, the inability of Vail to secure details about Henry's discoveries, or Morse's strictures about revealing trade secrets has been the subject of interpretation. Henry never deigned to answer Vail's letter.

Morse attempted again to heal the rift. He met Henry that winter but later claimed that Henry exploded at the very mention of Vail's name. "I will have nothing to do with Mr. Vail! What right has Mr. Vail to write the history of electromagnetism? He knows nothing about it!" Morse nevertheless promised that the errors in Vail's book would be rectified in the second edition. They were not, and Henry's fury was complete. He accepted an offer by a rival telegraph firm to act as a consultant—for remuneration—the first of several offers he would accept. Morse regarded this as treachery.

Relations were severed, but the fight for posterity had only begun. Under subpoena, Henry testified on Morse's behalf in four trials in the late 1840s, including *Morse vs. O'Reilly*. The substance of the testimonies was that Morse "had made no discoveries in science [but that] he was entitled to the merit of combining and applying the discoveries of others, in the invention of the best practical form of the magnetic telegraph." Furthermore, Morse was "not entitled to the exclusive use of the electro-magnet for telegraphic purposes, he was entitled to his particular machine."

Morse, damned by faint praise, attempted to return the favor. At a meeting with Sears Walker, who was authoring a report for the Coastal Survey on the history of the telegraph, Morse initially claimed not to have met Henry until 1847!* Gale, also present, corrected him. Later Morse sent Walker a paragraph parsing credit as he saw it due: "I claim to be the first to propose the use of the electro-magnet for telegraphic purposes, and the first to construct a telegraph on the basis of the electromagnet, yet to Professor Henry is unquestionably due the honor of the discovery of a fact in science which proves the practicability of exciting magnetism through a long coil or at a distance, either to deflect a needle or to magnetize soft iron." One can read this as proper acknowledgment of a scientific discovery, but certainly Morse did not go out of his way to emphasize that this discovery was precisely the basis for the telegraph. Walker chose not to include the statement in his report.

That was the beginning. By 1854 *Scientific American* had asserted in regard to the telegraph that "our country is more indebted to Joseph Henry than any other living man, and he has neither received the public credit nor honor which are justly his due," a claim that was echoed in Europe. Wheatstone called Morse a lucky robber of Henry's work. Morse had had enough. In 1855 he penned "Defence Against the Injurious Deductions Drawn from the Deposition of Prof. Joseph Henry," which appeared in Shaffner's *Telegraph Companion* magazine. It is a remarkable document, difficult even now to stomach for its sheer viciousness. In intensity it rivals the earlier *Conspiracy,* and one can only suspect that the criticism he received had again unleashed the darker side of Morse's personality. For *ninety pages* he assails not "injurious deductions" but Henry himself, attempting in every way possible to discredit his morals and scientific reputation. It becomes "not less a duty to the cause of Historical truth . . . to expose, as I shall do, the utter non-reliability of Prof. Henry's

* The same Walker who appeared in the Neptune saga (chapter 4).

testimony." He goes on to claim he derived from Prof. Henry "no aid in personal interview or by correspondence" and denies even that Henry's "alleged discoveries" had any influence on him. Those "alleged discoveries," furthermore, had no bearing on the telegraph whatsoever. Throughout, Morse rearranges dates—as he claims Henry has done—asserting that by 1835 his invention was essentially complete and that he and Henry did not meet until 1839. Any sympathy one had for Morse's position is largely dispelled by this document. One can summon up only a distant pity.

So deep were the wounds inflicted on Henry by Morse's article that Henry asked the board of the Smithsonian to convene an inquiry into the matter. It did, and in its report of 1858 it soundly condemned Morse, exonerating Henry on all particulars. Given that he was secretary of the Smithsonian, it might seem that the board had little choice. Morse retaliated by reprinting his article as a book.

There it roughly ended. Morse became ever more famous. Not a secessionist, he found biblical justifications for slavery. He helped lay the first transatlantic telegraph cable, which broke. He and his English rival Cooke were reconciled; he and Joseph Henry never were. Henry refused to attend Morse's funeral, which took place in 1872. Congress held a memorial service in his honor.

More than the other tales in the Panopticon, the story of the telegraph raises the question of what it means to invent. Morse did not invent the telegraph, though he claimed to have invented the electromagnetic telegraph. I seriously doubt Henry would have claimed to have fathered a practical telegraph, though his discoveries with electromagnets were crucial for its practical success. We do not know exactly what Henry said to Morse during their visits, or how much was transmitted to Morse by others, and later the ill will had reached such a pitch that all reconstructions became unreliable. The fact is, Henry did not patent his discoveries, and much grief might have been avoided if he had. To Morse must go all credit for perseverance through thick and thin to get his system established. His was the simplest and most robust of all the telegraphs, and so it quickly displaced its rivals.

Morse was undoubtedly aided by his ignorance. He had heard of electromagnets in the Athenaeum lectures and put his single piece of knowledge to good use. Had he a broader knowledge of the research of the times, he undoubtedly would have designed something more sophisticated and less reliable. He was also aided by a blindness characteristic of single-minded men. "God has chosen me as the instrument," he frequently claimed, "and given me the honor." In the conviction that he was an instrument of God, Morse knew he was absolutely right. It is sad that his rightness did not engender enough Christian charity to acknowledge others. The telegraph was in all probability the most important invention of the nineteenth century, and one of the greatest of all time, but we may wonder whether even that is worth a friendship.

12 / Fiat Lux: *Edison, the Incandescent Bulb, and a Few Other Matters*

Writers should be grateful for generational memory. Every twenty years a new generation appears with no knowledge of the past and a certainty that the past cannot be of any importance. In some vain attempt to perpetuate a historical sense, publishers hire authors to retell stories that have been told before, submerged, and told again in the unending cosmic cycle of forgetfulness and remembrance. The stories are always the same, with perhaps updated language and ever more tantalizing graphics, yet without this unvarying wheel of extinction and rebirth, writers would find themselves with much free time on their hands and nothing by way of compensation. Knowledge, Plato understood, is merely recollection.

And so nothing I am going to say in this essay and the next has not been said before. Yet what interests me is whether tales in the Panopticon of Present and Past Concepts evolve. More than a few demythologizers, debunkers, and even television documentaries have pointed out that Edison was not only a great inventor but at times an unscrupulous—if not ruthless—competitor, and that his lightbulb was far from the first. Sufficient environmental pressure has been placed on the standard tellings that some Darwinian diversification and natural selection should have taken place. In terms of serious biographies, I believe this has occurred. If we glance back to 1940 and the Hollywood biopic *Edison the Man*, we can be beguiled by Spencer Tracy's utterly ingratiating performance of a man almost devoid of human qualities. The inspiring—and it is inspiring—portrait of a pixieish, warmhearted fellow who is fond of children and selflessly devoted to the betterment of humanity is as seductive as it is iconographic. Watching the film, you'd never guess that Edison's first marriage was a disaster and that his wife died in a state of mental collapse, that he was aware of a single one of his predecessors, that he did anything but reject financial backing in his quest for the incandescent bulb, or that any of the Menlo Park crew other than himself ever had an idea. Traveling back even further to 1931, Francis Trevelyan's *Thomas A. Edison, Benefactor of Mankind; the Romantic Life Story of the World's Greatest Inventor*—well, I think the title says enough.

Contrast those portraits, if you will, with this passage from Neil Baldwin's 1995 biography *Edison, Inventing the Century.* We are in the midst of the infamous "current wars." Despite the fact that transmission of AC power is far more practical than transmission of DC, and despite the fact that one of Edison's own engineers, Nikola Tesla, had invented the AC motor, Edison remained adamantly, intractably opposed to AC and its foremost proponent, George Westinghouse:

> For two years after Tesla deserted him, Edison spent increased time and energy in a morbid public relations campaign to demonstrate that the compelling reason to shun alternating current was because it was dangerous and direct current was safe. In the courtyard of the West Orange Laboratory, media-event public executions of animals by alternating current were conducted, escalating from dogs . . . to calves to horses.

Edison's attorneys went on to suggest that Westinghouse's name be used as a synonym for "to electrocute criminals with alternating current." Edison surreptitiously managed to have the first electric chair installed in Sing Sing prison in order that the public would be horrified by AC and bring Westinghouse's death knell. He was wrong on both counts.

We conclude evolution has occurred. Yet, as neo-Darwinists understand, evolution does not imply progress. When we visit the National Park Service's Edison Web site, we find the banner header "Edison's Top Three Inventions," which are of course the lightbulb, the phonograph, and the motion picture camera. Buried in deeper recesses is an acknowledgment that Edison didn't truly invent the incandescent bulb; he invented the first practical one. Even this modest concession, the visitor feels, has been made at great psychological cost, for one searches the site in vain for the name of a single one of Edison's precursors. Nor will one find a concession that maybe he didn't entirely invent the motion picture camera either. Whether he did or he didn't is not as fascinating as the tenacity with which the curators of the exhibit cling to not even a portrait but an abstraction, much like Spencer Tracy's Edison sixty years removed.

Thus at the Panopticon great counterforces are at work to stabilize traditional accounts. So genetically ingrained in every one of us are the heroic portraits, that the first reaction of Panopticon curators when faced with a discordant fact is to dismiss it out of hand as erroneous. If the anomalous information persists, a curator's next reaction is to acknowledge the correctness of the information but diminish its importance. Ultimately a kind of Orwellian compartmentalization is reached in which the new facts are accommodated but the overall picture lives on unaltered. It is much the same as when you confront a scientist with any new result. One of three reactions is inevitable: "It's wrong," "It's trivial," or "I did it first."

So Edison didn't invent the lightbulb, but he could say, "I did it better." There were many precursors, at least two dozen by my count, some famous and some who might as well never have existed, and Edison certainly knew of their work; the first assignment of Francis Upton, hired by Edison on the eve of the lightbulb push, was to do a patent search. Histories of Edison's bulb tend to dismiss the earlier attempts as primitive failures. "The simple fact" write Robert Friedel and Paul Israel in *Edison's Electric Light,* "is that the world had nothing even resembling a practical electric lamp. . . . It is not right either to make a great deal of the rivals Edison met in the field, whether in America or overseas, and to see them as equals in the enterprise. The evidence is simply not there to support the claim that any of these men possessed more than a portion of the whole that emerged from Menlo Park." Well, judge for yourself.

Before the shadows, though, Menlo Park. The historic search that produced the light at the end of the tunnel has itself become part of American folklore: thousands upon thousands of substances tried over the period of a year until on that immortal day in October 1879 Edison lit upon a simple cotton thread impregnated with carbon. How Edison himself got the idea for carbon filaments is an enduring mystery. One can see the screenwriters for *Edison the Man* poring over this description of the famous discovery:

> Toying one night with a piece of lamp-black mixed with tar (used in his telephone), he noticed, as he rolled it between his thumb and forefinger, that it readily became elongated, and the thought struck him that a spiral made of it might emit a good light under the action of the electric current.

One feels that this description was one of those leaked to the newspapers by Edison himself. There is in fact no documentary evidence that sheds light on when Edison's original experiments with carbon began. Both Edison and Charles Batchelor, Edison's most trusted collaborator (in the movie version, the guy with the big beard), testified in court that they had begun experiments with carbon strips in 1877, then put them aside when they proved unsatisfactory. Two years later, during the breakthough month, Edison was preoccupied with his improvements to the telephone (next exhibit) and seems to have participated little in the work on the incandescent bulb. It is telling that in early books on the electric light, the personal pronoun *he* is omnipresent to the extent that a reader might easily forget Edison had help, yet it is Batchelor's notebooks that are filled with details of Batchelor's own attempts to make carbon filaments. Lampblack was everywhere at the lab due to its use in the carbon-button transmitter for Edison's telephone (see the next chapter), and the historic decision to turn to carbon may have been simply one of expedience. The famous Number 9 of October 22, which burned for 13½ hours, was merely the beginning of the experimental run. Day by day, Batchelor and Upton tried variation and variation, until the series went up to 260. Edison had already applied for the master patent on November 4.

The Canadians tell a somewhat different version of this story. According to a detailed article in *Canadian Electrical News* from 1900, a medical student named Henry Woodward and a hotel manager, Matthew Evans, produced a workable carbon lamp within a nitrogen atmosphere in 1873, receiving a Canadian patent in 1874 and a U.S. patent in 1876. The last is certainly true: The patent is number 181,613, and you can obtain it online from the U.S. Patent Office. Woodward and Evans' first lamp was made from a glass tube into which a piece of carbon was hermetically sealed; the tube was evacuated and filled with nitrogen (if he hadn't done this, Woodward might have become world-famous). Judging from the patent and description, it seems as if the working element was not a carbon filament (Edison's crucial advance) but a thicker wire or even a rod. Nevertheless, as Evans described the trials with the initial six lamps:

> There were four or five of us sitting around a large table, Woodward at the head. The six lamps were strung in series from two supports on the table. Woodward closed the switch and gradually we saw the carbon become first red and gradually lighter and lighter in color until it beamed forth in its beautiful light.

Evans doesn't tell us the date of the trial (probably 1874) or how long the lamps lasted, but the success was sufficient to attract a number of backers. One of them, an A. M. Sutherland, recounts that Woodward ran into an interference suit with a Russian "whose name I think was Ladigan." (Lodygin, of whom more shortly.) Although Woodward won the interference suit, legal expenses were becoming great and progress slow, and the backers pulled out. So disheartened was Woodward that he emigrated to England, never to be heard from again.

But the postscript is the most interesting. Sutherland continues:

> A few years after this I moved to New York, and in the course of business I was thrown into more or less intimate contact with the now famous electrician, Thomas A. Edison, who was at that time experimenting with the electric incandescent lamp at Menlo Park, N.J. In talking about it one day, I chaffingly said: "Why Edison, you are nowhere! I am part owner of a patent several years older than yours on about the same thing you claim." He asked what patent I referred to and I told him, he remarking: "O, I know that patent, it is no good." The next day one of Mr. Edison's men, whom I knew very well, asked me how much I would take for my claim in the Woodward Patent, and that day I sold it to this gentleman, who in turn made over the papers to Mr. Edison. This is all.

This, of course, is not all. Since 1802, when Humphrey Davy at the Royal Institution (probably in the same lecture hall as Thomas Young) ran an electric current run through a platinum wire or carbon rod and demonstrated that it would glow until consumed by oxidation, investigators understood that to make an incandescent bulb one required a substance that would not burn up. Alternatively, one might create an environment from which the oxygen had been removed. Woodward and Evans chose a nitrogen atmosphere as one in which carbon would not burn. But theirs was the latest in a long tradition. Reports indicate that as early as 1820 the English experimenter William de la Rue attempted to create an incandescent lamp by placing a platinum wire in a partially evacuated glass tube. Presumably we would have heard had there been an astonishing success. Most of the several dozen others who made earlier attempts devised similar strategies: placing a piece of platinum or carbon in either a nitrogen atmosphere or a vacuum.

Vacuum is the key word. The birth of a practical incandescent lamp necessarily awaited the development of vacuum pumps capable of creating about a millionth of an atmosphere of pressure. These were the first mercury diffusion pumps, which were developed only in the 1860s (notably by Geissler and Sprengel), and one might plausibly argue that this was the true breakthrough. Edison made good use of it, borrowing a mercury pump from nearby Princeton University and having new ones made by his glassblower at Menlo Park.

The other major component of an electric lighting system is electricity. Without a scheme of generating and distributing current, an isolated bulb is more a novelty than a revolution. And now we perceive why the phenomenon of electromagnetic induction, termed in an earlier exhibit the basis of civilization, is. Joseph Henry and Michael Faraday (or vice versa) discovered that a magnet moving through a coil of wire produces an electric current. Henry also constructed the world's strongest electromagnets by increasing the number of turns around the iron. An electromagnet of many turns rotating between the poles of a permanent magnet is nothing less than a generator or dynamo. In 1832, within a year or so of Faraday and Henry's experiments on induction, a Frenchman, Hippolyte Pixii, constructed a tabletop dynamo, for which he received a patent. It might have been able to power a flashlight bulb. Exactly thirty years later a much larger dynamo supplied current for the first commercial use of an electric light—an arc lamp in the Dungeness Lighthouse in England.

Once these ingredients were in place, the race for the incandescent bulb heated up. In Russia, Aleksandr Lodygin, with whom Woodward crossed paths, is regarded as the inventor of the incandescent bulb. Lodygin (in older books Lodyguine) received a Russian patent in 1872 for a bulb with a piece of carbon housed in a nitrogen atmosphere. Judging from the specifications of Lodygin's bulb, his carbon, like Woodward's, was closer to a rod than a filament, which meant it had low resistance and so would not burn brightly and would require high current. This is not so surprising. Ohm's law—the basic relationship among electrical resistance, voltage, and current—was not clear to the investigators of the day, and many of them didn't seem to realize a thin filament of high resistance would produce the most light at the least risk of burning out. Nevertheless, Lodygin's lamp was successful enough that he installed about two hundred of them at the Admiralty Dockyard in St. Petersburg as an experiment. Two years later the Academy of Sciences awarded Lodygin the Lomonosov Prize, one thousand rubles, and he formed a company. According to Western sources, the lamp proved too expensive to operate (presumably because of high current), and the endeavor failed. An unpublished manuscript by one of his children in Columbia University's Rare Book and Manuscript Library tells a somewhat different story: The investors, in an action that will sound familiar to today's public, made inflated claims for the lamp, which sent the stock price soaring and Lodygin scurrying nightly to his lab to catch up to the publicity. A few months of this left Lodygin "a nervous wreck." Enough was enough. Lodygin "refused categorically to have his name attached with shady speculation and broke up the company by leaving it." Then, in a Tolstoyesque move, he decided to establish a model farm in the Caucasus and went off to the mountains to become a peasant.

A few other surprises are found in the Lodygin file. His nitrogen bulb is barely mentioned. Lodygin also developed an evacuated bulb, though because he used a hand pump and even a mouth pump, the quality of the vacuum must have been hopelessly inadequate. His first lamps burned for about twelve hours, but after some time—evidently around 1880—Lodygin extended the lifetime to upward of a thousand hours. Later, in 1890, he would construct bulbs with metallic filaments, including tungsten, though I have not been able to ascertain how successful they were. Most surprising of all is that, according to a rather detailed series of articles by Albert Parry writing in *Novoe Russkoye Slovo,* not only were Lodygin's lamps successful enough to have been used underwater in 1875 during work on one of St. Petersburg's famous bridges, but a naval engineer, Lieutenant A. M. Khotinsky, came to America in 1877 and showed Lodygin's bulb to Edison himself. We imagine the ensuing conversation was similar to the one Edison had with the Canadian Sutherland.

Mysterious military men engaged in even more mysterious intrigues are a staple of Russian history. On the other hand, history is made in the interstices of the written records. Those who believe that destiny is woven by the stars will find it unavoidable that Lodygin, born in the same year as Edison, finally emigrated to the United States to work for Edison's archrival Westinghouse, then Sperry Gyroscope. He died in 1923.

Russians notwithstanding, the most famous of Edison's competitors was undoubtedly Joseph Swan, a respected English inventor who had been experimenting on and off with an incandescent bulb since about 1848. In 1877, one year before Edison himself stated that he had begun work on carbon filaments, Swan was doing the same, and in an evacuated bulb. In 1879, eight months before Edison and Batchelor's breakthrough, Swan demonstrated his lamp at a meeting of the Literary and Philosophical Society in Newcastle. Swan even wrote to Edison when informed of his rival's success:

> I think I am in advance of you in several points especially in the making of the carbons. . . . I can easily convince you if necessary that I have been working a long time on this subject and that carbonised cardboard was a material that I have for years been experimenting with and was actually working at the very time you announced your use of it. . . . I therefore had the mortification one fine morning . . . of finding you on my track and in several particulars ahead of me—but now I think I have shot ahead of you.

Though the letter is certainly polite, Swan apparently had second thoughts and never sent it, but his work was famous enough throughout England that Edison's breakthrough received a Gilbert-and-Sullivan-style parody in a London paper (to the tune of "When I Put This Uniform On"):

I said when I turned the Lamp on,
It's plain to the veriest dunce,
That all gas shareholders
Will shrug up their shoulders,
And sell off their gas shares at once.
I thought I was first in the Field,
And this lamp a good income would yield;
But my hopes are all blighted,
Completely benighted,
By a Briton whose plans were concealed,
Which I never counted upon,
When I first turned this little lamp on.

Well, the author got it wrong. Swan was more retiring than Edison, had less money, and, as might be gathered from the fact that he never sent his letter, was slower to act, especially in the matter of patents. All this was fatal when dealing with the American, who "believed that the race was not to the swift, but rather to the man with bravado, endurance and money." And so when Swan finally did patent, Edison claimed that the Englishman's lamp infringed on *his* invention. Litigation ensued, though happily (we assume) the matter was settled out of court and in 1883 the rivals formed the Edison and Swan United Electric Company, which until 1893 held a monopoly on manufacture in England.

There is a good lesson here for the editors of *Time* magazine and their profound deliberations over the inventor of television. Swan did it before Edison, true, but ultimately that's not the only question or likely even the most important one. As the editors of *La Lumière Electrique* put it, "Swan's lamp may have come before Edison's, but nothing presently shows that it is superior." Edison's main contribution was that he managed to create an entire lighting *system*. Back in 1878 Grosvenor Lowrey, the general counsel for Western Electric, had assured Edison that if he could develop a practical electric light, fortune would be his. One can see the wheels turning; Edison gave Lowrey the required assurances, and in the blink of an eye Lowrey had incorporated the Edison Electric Light Company, enlisting such stockholders as J. P. Morgan in order to corner the market on the production and distribution of electric light. But understand: This was before Edison had done any experiments whatsoever. That is what Swan was up against.

Edison had been releasing statements a year ahead of time that the lightbulb was already in hand (interestingly, in the movie the reports are leaked by an untrustworthy friend), and when success finally came it was greeted by disbelief, a full-page article in the *New York Herald,* skyrocketing shares in the Edison Electric Light Company, a drop in the price of gas shares, and jubilation as hordes of people made their way to Menlo Park to behold the miracle for themselves. Edison was already thinking of the future. He understood that for commercial purposes bulbs could not be connected in series (since if one failed, the whole circuit would go dark) but must be connected in parallel. This required a constant-voltage dynamo, which many thought impossible, yet Edison created one, and within four months the first commercial installation was made aboard the steamship *Columbia.* The famous Pearl Street Station in New York went online September 4, 1882.

Evaluating Edison's genius is no simple task. Much perspiration, for sure. An artist's appreciation for money and capital, market forces. A not-insignificant dollop of P. T. Barnum, a big dose of ruthlessness. And the help of a large and talented cadre of assistants. I am inevitably reminded of a chart of "the world's twenty most productive scientists" I saw posted on a bulletin board about ten years ago. The first column listed the names, the second column listed the number of papers published between 1980 and 1990, and the last column listed the days between papers. Most of the winners were heads of large biomedical or chemical laboratories. I believe Robert Gallo, the HIV investigator, was the top-scoring American. The winner had published something like a thousand papers in ten years, which amounted to one paper every 3.6 days. Now, nobody takes this sort of publication rate seriously. The only thing it can mean is that the author never saw 99 percent of them. While one can and usually should give a lab director credit for the overall direction of the research team, it would be silly to pretend that he is responsible for everything that goes on in the building.

It may be too much to view Edison as the first of the modern laboratory managers, but one can't deny he often merely outgunned his poorer competitors. The Frenchman Charles Cros, for example, designed a workable phonograph slightly before Edison, but Cros was an impoverished teacher (and great poet) and could not even afford to apply for a patent, so he deposited signed and sealed plans with the French Academy of Sciences.* For all this, though, one can still be impressed—if not amazed—at the fecundity of Edison's imagination. If Edison was often preceded by others, his version was invariably better.

Clearly one reason Edison looms so large in the American imagination is that he fits our culture's folk image of its heroes—a self-made man who spurned too much education. (His agnosticism, if not atheism, has been toned down for public consumption.) Edison's achievements seem within the realm of the possible—hard work and no higher mathematics.

Edison had more than a little in common with Roald Amundsen, who beat Robert Scott to the South Pole. Scott's expedition was poorly organized; they had taken horses and tractors instead of dogs and sleds and as a result performed prodigies of heroism and died. Amundsen, on the other hand, displayed a certain sagacity at getting to the Pole. He decided on dogs, guessed rightly at a shortcut, and made it to the Pole and back with no muss and no fuss. Scott became famous for the tragedy, but in the words of one of the expedition members, "Tragedy wasn't our business." Amundsen got the job done. He may not have been a nice guy, but getting the job done is the business of the inventor, and in that Edison had no equal.

---

* Cros ends his description by saying, "In any case a helical trace on a cylinder is much preferable, and I am presently occupied in trying to find a practical realization." Earlier, in the *Comic History of the States and Empires of the Moon* of 1656, Cyrano de Bergerac predicted the phonograph: "He places the needle on the chapter to be heard and, at the same time, there come, as from the mouth of a man or from an instrument of music, all those clear and separate sounds which make up the Lunarians' tongue." In 1855 yet another Frenchman, Léon Scott, devised the "phonoautograph," a device to record sound on a cylinder covered with soot. However, there was no way to replay the sounds.

13 / "Magna Est Veritas et Praevalet": *The Telephone*

Technically, or at least chronologically, this story might well have come before the previous one. Edison invented his lightbulb in 1879, Bell his telephone in 1876. The telephone was a direct outgrowth of the telegraph and so should properly be set immediately in its wake, but it was also the immediate predecessor to the radio and so should be placed immediately before that. From either point of view the lightbulb interrupts the logical flow of events and so never should have happened.

But taking up the thread of the last chapter, that of portraiture—evolving or otherwise—I chose to place Edison first because he seems to me more "real" than Alexander Graham Bell, whose name every American associates with a corporation rather than a human being. Fewer people, I am sure, have read biographies of Bell than Edison, the National Park Service maintains no Bell Web site, and Google tosses out 50 percent fewer links for Bell than for Edison. The classic Hollywood biopic *The Story of Alexander Graham Bell,* in which Don Ameche plays Bell, is surprisingly accurate on technical details, if more creative in other matters, and in one or two places moving to tears. Yet despite the fact that the telephone briefly became known as "the Ameche" after the film's 1939 release, the film is impossible to find in local video stores.*

What is curious-strange is that, in contrast to Edison, Bell's portrait might well nigh have been completed with the movie. As the man who received "the most valuable patent ever issued," Alexander Graham Bell has undeniably passed from mortal to icon, and as befits an icon of the true orthodox church, the portrait has become static—no, ossified. This must be due in part to the fact that many of his biographies are written by either relatives or fellow Scotsmen, which is much the same thing, and certainly they have relied to an unconscionable extent on information supplied by the family or the Bell archives. In fact, once the reader gets over the nearly overwhelming impulse to hurl them all into the flames, he or

---

* It is available from some sources under the title *Alexander Graham Bell.*

she is left firmly and forevermore with the conviction that science would be better off without scientific biographies altogether. When you get down to it, biographies are much like patents: By their very nature they exclude everyone but the awardee, relegating runners-up to the outer darkness while illuminating the winner with a light too intense for mere mortals.

The telephone is history's most extreme case of patent exclusion. The invention was, I read, the subject of some 600 lawsuits that fill 150 volumes. Whether the priority arguments eclipse or merely rival those surrounding the radio, I don't know. I do know that a man's stock of hubris—maybe ignorance—would need to exceed all natural bounds to think he could sort out so many cases that may have been resolved legally but almost certainly have no resolution of any other sort. So I nobly choose to step aside from the fray. More informative of human nature is to focus on what the various sources say about their respective protagonists—and what they don't say. It is indeed what goes unsaid that snaps our attention to, and generates gas. Every author is limited by time and resources, but the air surrounding the telephone controversies is tainted with the unpleasant whiff of what has been soundly dropped by parties on all sides.

Let's kick off with a few of the Bell biographies. Anyone reading them with a view to understanding technological history, rather then Bell's history, will notice a *slight* tendency to diminish the work of the great man's predecessors. Take, for example, Robert Bruce's *Alexander Graham Bell and the Conquest of Solitude,* the most authoritative Bell biography. From it we learn of Johann Philipp Reis, who attempted electrical voice transmission well over a decade before Bell:

> Since Reis' instrument could not reproduce amplitudes or degrees of loudness, it could never transmit the subtle compound of many amplitudes that constitutes speech. It could convey the rhythms and pitches of speech and thereby the illusion of intelligibility—but only the illusion. Reis did not choose to patent what he regarded as a purely scientific experiment. And indeed it had no conceivable commercial use then, nor would it ever.

Edwin Grosvenor, a great-grandson of Bell, and Morgan Wesson published *Alexander Graham Bell: The Life and Times of the Man Who Invented the Telephone* in 1997. We read:

> In 1861, for example, a German schoolmaster named Philipp Reis perfected the transmission of electric tones, though not actual voice trans-

mission. Reis' description of his invention carefully stepped over the subject of intelligible speech.

In James Mackay's biography of Bell, also from 1997, we learn that Charles Wheatstone of England invented the telegraph, that John Logie Baird of Scotland invented the television, and that

> as it could not reproduce amplitudes or degrees of loudness, the Reis instrument could never have transmitted the subtle and extremely complex mixture of frequencies and amplitudes that constitute the sounds of the human voice. It could simulate the rhythms and pitches of speech, producing a stream of gibberish suggestive of speech, but no more than that. Reis himself was aware of the shortcomings of his invention and never patented it, regarding it as nothing more than a scientific experiment without practical application.

And finally, from the urtext itself, the autobiography of Thomas A. Watson:

> No one had ever made [the Reis telephone] talk. It could not, because a make-and-break current can carry only the pitch of a sound and none of its delicate overtones.

One might reasonably want to know what Reis would have to say about all this. Luckily he tells us. But who was the German schoolmaster? Johann Philipp Reis was born at Gelnhausen in 1834 and, as he once said, lived a life of "labor and sorrow." Raised by his father, a baker, and his grandmother after his mother died, Reis developed an early taste for languages and the natural sciences; he took private lessons in physics and mathematics, joined the Physical Society of Frankfurt am Main, became a secondary-school teacher at the age of twenty, married at the age of twenty-five, and died at the age of forty of tuberculosis. How or when he got interested in electromagnetism is not known, but as Reis himself wrote, "Incited thereto by my lessons in Physics in the year 1860, I attacked a work begun much earlier concerning the organs of hearing and soon had the joy to see my pains rewarded with success, since I succeeded in inventing an apparatus, by which it is possible to make clear and evident the functions of the organs of hearing, but with which also one can reproduce tones of all kinds at any desired distance by means of the galvanic current. I named the instrument 'Telephon.'"*

---

* Reis was not the first to use the term.

Over the next few years, Reis devised ten forms of a telephone transmitter and four of a receiver. It seems like an awful lot of work for something he never intended to function, yet after reading the Bell biographers' declarations one could be forgiven for believing exactly that. The idea that Reis never intended for his telephone to transmit speech—and that it couldn't even if he wanted it to—got started immediately after Bell patented his own telephone. Such was the denial that Reis might have preceded Bell that the English physicist Silvanus Thompson was able to write as early as 1883: "To all such clap-trap as this—and there has been enough *ad nauseam* of such—the one reply is silence, and a mute appeal to the original writings of Reis and his contemporaries." That seems an eminently sound approach.

Thompson (one of those who had pointed out that Becquerel's discovery of radioactivity had paralleled Niepce's forty years earlier and who will prove in the upcoming exhibit that he was never one to shrink from controversy) was good enough to collect and present some of Reis' papers. We find the following in Reis' 1865 prospectus to accompany his telephones sold by Wilhelm Albert of Frankfurt, "Besides the human voice (according to my experience) there also can be reproduced the tones of good organ-pipes from F to c̄ and those of a piano." He goes on to say that by use of his receiver, "it is possible to make oneself understood quite well and certainly by the other party" and suggests signaling in advance whether one tends to sing or speak.

This and other documents make it quite clear that Reis intended his telephone to work. Did it? Unless one could refurbish and test any of his instruments that might be extant, no one alive can state preemptorily whether his telephone was capable of transmitting speech. Thompson himself tested every version of Reis' telephone he could lay his hands on and, provided they were set up properly, "found them perfectly competent to transmit speech." Like all good partisans, Thompson tended to be rabidly biased in any cause he took up, but eyewitness reports are many. To cite just one, after the Frankfurt Physical Society meeting of October 26, 1861, when Reis demonstrated his telephone, members of the audience stayed behind to try it. "The singing was heard much better than the playing," reports Heinrich Friedrich Peter, who was there. Afterward, Peter tells us, he

> went up into the room where stood the telephone, and purposely uttered
> some nonsensical sentences, for instance. . . . "The sun is made of cop-
> per," which Reis understood as "The sun is made of sugar"; "The horse
> eats no cucumber salad," which Reis understood as "The horse eats. . . ."

Those who were present were very greatly astonished, and were convinced that Reis's invention had opened out a great future.

You'd be correct to surmise from this report that the telephone's performance was inconsistent, but Reis never claimed for it anything other than what he could demonstrate. The royal Prussian telegraph inspector did write in 1862, "That which has here been spoken of will still require considerable improvement," and in 1876 Bell himself said of his own patented telephone, "The effects were not sufficiently distinct to admit of sustained conversation through the wire. Indeed, as a general rule, the articulation was unintelligible . . . excepting when familiar sentences were employed."

Reis, Robert Bruce and James Mackay are quick to point out, never patented his telephone. They are loudly mute on the fact that he demonstrated it publicly at lectures all over Germany between 1861 and 1864, and as early as 1862 submitted a paper to the *Annalen der Physik*. Professor Poggendorff, the publisher, rejected it out of hand. The Physical Society paid no further attention to the telephone after the 1861 demonstration, and as a result, Reis resigned his membership six years later. By that time, though, his devices were being manufactured and shipped all over the world, including, not insignificantly, to the Smithsonian Institution. At some point Poggendorff decided the telephone was attracting enough attention to be written up in the *Annalen,* and he invited Reis to do so. Loosely translated into the modern vernacular, Reis told Poggendorf to go to hell. After around 1872 Reis' decline in health from pulmonary consumption precluded further work on invention. Before he died, Reis donated his instruments to the Garnier Institute for Boys, where he taught. For a few years he was forgotten, until Bell patented his telephone; then colleagues at Garnier erected a memorial obelisk to their friend. It was too late, of course, but in some act of atonement Germans to this day regard Reis as the inventor of the telephone. A postage stamp was issued in Germany in 1961 to honor him.

To follow the to-ing and fro-ing shortly to commence, it's important to have some idea of how Reis' telephone operated. Although his transmitters evolved through ten forms, they were all based on the same principle: mimicking the human ear. Reis' first version was in fact a carved wooden model of an ear with a thin membrane stretched across one end of a hole where the eardrum would be. Attached to the other side of the membrane was a small metal lever. When someone spoke into the ear, the membrane vibrated and the tiny lever rattled against a second piece of metal, creating a loose contact. The idea of regulating electric current by loose, imperfect connections is the idea behind David Hughes' microphones of chapter 6 and the "coherer," to make its appearance soon. In Reis' telephone, the ear "microphone" controlled the size of the current sent to the receiver. The tenth version was quite different in appearance. The speaker talked into a tube whose other end had been fitted into a sound box. Stretched across the top of the box was a membrane that caused one small piece of metal to vibrate against another. All of Reis' transmitters employed the idea of a loose connection set oscillating by a membrane. The tympanum, as he referred to it, featured in all subsequent telephones.

Reis' receiver was a different matter. Three of the four versions employed a knitting needle around which was wound a coil of wire. As the oscillating current from the transmitter passed through the coil, the needle actually emitted sounds. (It's unlikely Reis understood why, but it certainly had something to do with the rapid magnetization and demagnetization of the needle.) Reis amplified the sounds by use of a resonance box. In his Mark 1, he actually rested one end of the knitting needle on the sounding board of a violin; in his later models the needle was cradled horizontally over the box. For a brief time Reis took a detour with a receiver consisting of a electromagnet that vibrated with the incoming current against a flexible armature or flange, which was in turn connected to the sounding box. This arrangement evidently gave unsatisfactory results, and Reis quickly abandoned it, reverting to the knitting needle.

One would be forgetful indeed to conclude from all this that Reis' ideas arose from a vacuum. The Infinite Chain of Priority asserts that Reis himself had precursors. The first person who clearly enunciated the requirements for electrical transmission of speech seems to have been a Frenchman, Charles Bourseul, who did so in 1854. In his lectures and papers Bourseul is quite specific, not to mention prescient, declaring that if one had a metal disc that would be flexible enough "to reproduce all the sound waves transmitted to it by the air," then the disc could start and stop an electric current that would cause a second diaphragm to vibrate in the same way. Bourseul slips through the cracks in Bruce's book, and according to Mackay he failed in his attempt to transmit speech (one somehow suspects Mackay wants to say "abysmally failed"), but the fact is that there is no record of any attempt being made. Whether Reis knew of Bourseul's ideas is unclear. On the other hand, Reis did apparently base his receiver on an earlier one of somebody called Noad, dating from 1857. By now you will be asking whether Bell knew of Reis' work. Yes, and he never denied it; more on that presently.

*   *   *

Let us now confront the main event. Specialists will forgive me for not discussing the work of Antonio Meucci and several dozen others whose names have sunk without a trace. To frame events, it might be useful at this stage to quickly dust off the portrait of the great frozen icon. Alexander Graham Bell was born the same year as Edison, 1847, in Edinburgh, Scotland, son of Eliza and Melville, a well-known elecutionist and teacher of the deaf. In 1870 Alec emigrated with his parents to Canada; soon he moved to Boston, where, despite the movie's romantic portrayal of a young man on the verge of starvation, he quickly arranged a busy life teaching the deaf, lecturing and inventing, and courting the daughter of his principal backer, Gardiner Hubbard. In 1875, with a little money from Hubbard, he hired Thomas Watson, a scientifically inclined machinist, and their first working telephone came about on March 10, 1876.

What made Bell's telephone different from its predecessors? His transmitter, like Reis', used a diaphragm set oscillating by the human voice. Like Reis', the membrane caused an electric current to flow from the transmitter to the receiver. But the crucial advance claimed by Bell was what he called an "undulatory" current, a current that varied smoothly as the membrane vibrated. Remember, attached to Reis' membrane was a little lever that rattled loosely against a second one. Either they were in contact or they weren't. In the parlance of the day this was referred to as a make-and-break connection.* Bell, in the version of his phone tested on March 10, employed a variable resistance to regulate the current smoothly to be able to transmit all the undulations of the human voice. In other versions the oscillating membrane, via the phenomenon of induction, would set up similar oscillations in a secondary circuit that were sent on to the receiver.

The warfare over the telephone—probably all 600 lawsuits—is almost entirely focused on whether this idea of undulatory currents was revolutionary, insubstantial, or even existed; whether Bell thought of it himself, pilfered the idea from his rivals, vice versa, or neither; and, if the last, whether Bell had the idea first, second, or third. Believe me when I say no consensus is to be found. About the only thing all parties agree on is that, yes, on that immortal March 10, 1876, Bell did shout something like, "Come here, Watson, I want you!" into his telephone, but the whole business about it being an accident caused when Bell spilled acid on his leg

---

* For more crucial and fascinating points about make-and-break connections, see endnote to this page.

was a yarn Watson spun for his autobiography, published fifty years later. Neither Bell nor Watson mentioned the incident at the time or afterward.

According to Robert Bruce, Bell got the idea for a diaphragm transmitter and receiver a year and a half earlier, in October 1874, and Bruce puzzles over Bell's failure to think in terms of a membrane before that date. It is truly puzzling, for on the previous page Bruce tells us that Bell was familiar with Reis' work, and many of Reis' diagrams show the membrane transmitter, including one published in Bruce's book. It was not until the following May that Bell got the idea of how to produce an undulatory current (by a variable resistance). In the meantime, in February or March 1875, Bell would see Reis' apparatus at the Smithsonian when Joseph Henry himself showed it to him.* Bell wrote home about that celebrated visit, in which he sought out the great scientist for advice about his work, Henry assuring Alec that he "had the germ of a great invention" and that he should get down to it. Bell answered that he felt he "had not the electrical knowledge necessary to overcome the difficulties." According to Bell, Henry's laconic reply was "Get it." Bell testified to all this in court, and there is no dishonor to it. Why there is dishonor to admitting he might have been influenced by Reis' ideas, I don't know.

For the briefest instant, let us leave the young inventor and his assistant as they struggle toward the telephone, and introduce the archrival. Elisha Gray, born in Ohio in 1835, is remembered now almost exclusively because he came in second in what is probably the greatest and most consequential coincidence in the history of technology. On the afternoon of February 14, 1876, Gray filed with the U.S. Patent Office in Washington, D.C., a caveat—a formal notice of work in progress—for a telephone. An hour or two earlier, at the same office, Bell had filed his own patent application. Despite his fate to be relegated to mean-spirited lists of also-rans, Gray was more famous than Bell at the time, being considered, after Joseph Henry, the most knowledgeable American about electrical matters and being the founder of what became Western Electric. Enthusiasts of musical history also know Gray (or maybe his nephew) as the discoverer of the singular singing bathtub, which birthed his musical telegraph, often described as the first "electronic" musical instrument.† In the early 1870s both Bell and Gray were involved in trying to develop "harmonic tele-

---

* For more on Bell and Reis, see endnote to this page.
† Gray or his nephew discovered that a metallic bathtub when stimulated by an electric current would emit sounds.

graphs," telegraphs that could transmit a number of messages simultaneously. (Edison's quadruplex was the most famous of these.) Their ideas for the telephone sprang directly from these efforts. By 1874 the two men had become aware of each other's work, and as 1876 approached, the rivalry turned into a desperate race to be the first to come up with a working instrument. To that extent, the great coincidence was not entirely coincidental.

Due to the virtually simultaneous filings, both claims were suspended pending review. Because Bell had technically come in first and Gray had filed only a caveat, the Patent Office decided in Bell's favor, and on March 7 he was officially awarded U.S. patent 174,465. The following fact is crucial: Bell submitted his patent application on February 14, but the cry for Watson did not take place till nearly a full month later, on March 10. The application itself, for "improvement in telegraphy," is rather broadly written, concerned almost entirely with the undulatory current idea. In terms of an actual device, by far the largest share—including the only diagram—concerns a device employing a membrane and electromagnets, the latter in a configuration almost identical to the one Reis had used for a short time. About fifteen lines are devoted to the possibility of a "liquid transmitter," the variable resistance device alluded to a moment ago. Here a wire attached to the diaphragm would dip into a conducting liquid under vibrations from the human voice. "The more deeply the conducting wire is immersed in the mercury or other liquid, the less resistance does the liquid offer to the passage of the current. Hence, the vibration of the conducting wire in mercury or other liquid included in the circuit occasions undulations in the current." Bell really doesn't say much more about it, and a few lines later he states, "I prefer to employ for this purpose an electro-magnet," referring to the diagram of the Reis-like instrument. One might also point out that the word *telephone* is nowhere to be found in the application, and although Bell mentions that the membrane of the device might be stimulated by the human voice, neither does he state anywhere in the patent that its purpose is for the transmission of human speech other than "telegraphically."

Believers in synchronicity will find much to stimulate their imaginations here. The first curiosity is that Gray's caveat was almost *entirely* devoted to the liquid transmitter, and he describes his ideas in much greater detail than Bell. The second curiosity is that Bell's immortal experiment of March 10 was made *not* with the electromagnetic device "preferred" in the patent but precisely with a liquid transmitter. (The sulfuric acid Bell didn't spill was to increase the conductivity of the liquid.) The third curiosity is that the entire paragraph describing the liquid transmitter was written in the margin of Bell's draft just before sending it to his attorney, as can be seen in the original at the Library of Congress. The fourth curiosity is that on February 26, between the filing date and "Watson!", the patent examiner Zenas Wilber revealed to Bell that the point of interference between his application and Gray's was precisely that paragraph about the liquid transmitter.

Now, wrongdoing on anybody's part, except maybe the patent officer's, has never been established (Bell did win all his lawsuits), but the spin given to these events by the various advocates is mighty interestin'. The Bell camp never fails to point out that the very fact Gray filed a caveat showed he hadn't built an apparatus at the time. "Gray had not put his ideas to any practical test," asserts Mackay unrepentantly. Can't deny it, but as demonstrated by the fact that Bell's first successful telephone was of a type not described by the body of the patent, Bell hadn't either. It's truly difficult to quote an absence in an exciting way, but as far as I can tell, the Bell biographers somehow manage to overlook this fact entirely.

Bellians, Mackay in particular, find the timing of Gray's caveat filing suspicious. He says, "What is remarkable is that Gray had never mentioned this telephone concept to a living soul. . . . In fairness to Gray, it is probable that some vague notion of a telephone had been forming in his mind for some time was brought into focus when he got wind of what Alec was about to patent." Interestingly, none of the biographers have ever come up with a plausible explanation for Bell's last-minute insertion of the paragraph on the liquid transmitter other than that he simply had a memory lapse, which they are quite happy to forgive. So am I. They certainly display no hint of suspicion that Bell might have got wind of Gray's work rather than the other way round. Mackay's tone is similarly nonchalant when he describes the patent examiner's revelation of February 26:

> During his meeting with Wilber, Alec casually asked him what aspect of Grey's caveat had caused the interference in his application. Theoretically, Wilber was bound by confidentiality not to reveal such details, but as the interference had been quashed he saw no harm in telling Alec that it was the matter of liquid variable-resistance.

So that you might judge for yourself whether Bell's interest was as casual as claimed, here is Bell's own deposition, given to the court:

> During this period I was not in Washington, but I arrived there on Saturday, February 26, and of course I was anxious to know what the point of interference had been. I therefore took the opportunity, while in the Patent Office on the business described in answer to Interrogatory 108, of asking the Examiner what the point of interference had been, and in reply he pointed to the variable-resistance clauses in the body of the

specification, declining to give any more definite information, as a caveat, he said, was a confidential document.

And for a third slant on history, here is how Lewis Coe describes the same events in his 1995 book *The Telephone and Its Several Inventors:*

> At the time [spring 1876?], Bell did not know how to construct a variable resistance transmitter, but the principle was vital to all the transmitters that were subsequently developed. The evidence suggests, indeed the patent officer at the time admitted, that he had shown Gray's caveat to Bell, enabling the latter to modify his application. This was out and out fraud, yet the courts disregarded the apparent facts and sustained Bell's claims. Gray was to realize later that he had apparently shown Bell how to construct the transmitter with which the first results were obtained.

One longs for history without interpretation and at the same time wonders what history would be like without it. The moral of the story is that it's better to file a patent application than a caveat, and you can be sure Gray was kicking himself for following the advice of his lawyers, who profited anyway. (Maybe Gray's misfortune is why caveats no longer exist.) Yet, contrary to popular belief, or at least as implied on the PBS Web site, Gray never took legal action against Bell. If he had challenged Bell's patent, there's a good chance many of us would have grown up cursing Ma Gray rather than Ma Bell. Gray did, though, appear as a witness for the defense in the most famous of the Bell lawsuits, the one Bell brought in 1878 against Peter Dowd and his employer, Western Union, for patent infringement. Relations between Gray and Bell had been cordial enough, each feeling the other had behaved honorably, but as a result of all he heard during the drawn-out legal battles, Gray eventually became convinced that he had been hoodwinked. Mackay, Bellian to the bitter end, describes it this way:

> Gray's chagrin was intensified when Zenas Wilber naively admitted to him that he had let Alec Bell know the substance of the Gray caveat. Wilber, in fact, went further and swore an affidavit asserting that he had allowed Alec to examine the caveat in detail, which was certainly untrue. It is possible that Wilber was induced to perjure himself by agents of the Globe Telephone Company.

The word *certainly* should be proscribed from historical discourse. Not only did Wilber swear an affidavit asserting that he had shown Bell Gray's caveat, but he swore one testifying that Bell had bribed him with one hundred dollars. Six months earlier Wilber had testified that no fraud of any sort surrounded the dual filings, and so this was a complete about-face. To make matters worse, Wilber had a history of alcoholism, which he acknowledged in his affidavit, and so there has always been doubt as to the veracity of his claims. Bell, it hardly needs saying, swore his own affidavit denying the whole thing. More than that only the Lord Almighty knows. Interestingly, Grosvenor and Wesson in their biography take no sides on this episode.

All this is mostly legal wranglin', but there is a moral issue centered on who really had the idea for variable resistance. The Bell camp points to a letter Alec wrote on May 4, 1875, in which he does (clearly, it seems to me) describe how an undulatory current might be produced by a variable resistance. However, as we've just seen, they deny Gray could have had a similar idea. "Not until late fall of 1875," writes Bruce, "did Gray, even by his own account, conceive the idea of a liquid variable-resistance transmitter."

Gray's supporters have a different story. "I believe," declared Frank Pope, "the discovery of the true method of transmitting composite vibrations was first publicly announced in the *Journal* [of the American Electrical Society] Vol. 1, also in Gray's patents, British No. 1,874/76 and U.S. 186,340." I have obtained a copy of the *Journal* and the latter patent. The patent, for improvements in electroharmonic telegraphy, does indeed assume an undulatory current, though it is not very clearly explained; in any case, the application was not filed until January 27, 1876, so its citation is irrelevant to the present discussion. As for the *Journal*, it contains several undated lectures by Gray that touch on "composite vibrations." The first, referring to experiments done as early as the winter of 1873–74, nowhere mentions transmission of voice. In the second Gray explicitly requires that a speaking telephone "copy with the greatest accuracy" the composite vibrations produced by the human voice (in other words, utilizes an undulatory current), but the talk refers to the 1876 patent in the past tense and so is also irrelevant to the discussion.*

When they're in a generous mood, Gray's advocates—rather, Bell's opponents—also point out that both men were anticipated by the ubiquitous Edison, with his patent 141,777, dating from March 1873, a patent that supposedly also describes a liquid variable-resistance mechanism. I've

---

* In his deposition for the Dowd case, Pope also refers to Gray's patent of July 27, 1875, for the transmission of musical tones (the argument being that the principles of transmitting music and speech were identical). This patent also embodied Gray's first receiver, which opens another can of worms, that of Gray's *receivers* versus Bell's *receivers*. See *Bruce vs. Taylor* in the references.

received patent 141,777 too, and while it definitely describes a mechanism almost identical to those of Gray and Bell, the device has nothing to do with the transmission of speech.

Such malcitations show that Gray's advocates are not lacking their own asperities. Most severely, the Gray camp routinely omits mention of the letter he wrote to Bell on May 5, 1877, and forgot about, but which was unearthed by Bell's wife, Mabel, during the Dowd case:

> Of course you have had no means of knowing what I had done in the matter of transmitting vocal sounds. When, however, you see the specification, you will see that the fundamental principles are contained therein. I do not, however, claim even the credit of inventing it, as I do not believe a mere description of an idea that has never been *reduced* to *practice*—in the *strict sense* of that phrase—should be dignified with the name invention.

At hearing this letter in court, they say, Gray turned to his counsel with an oath: "I'll swear to it, and you can swear at it!" But how much fire this smoking gun emits isn't as obvious as it might first appear. The letter was written before Gray had learned about Wilber and before he had become convinced that Bell really had stolen his ideas. Turnabout applies too. If we read in this letter Gray's confession that he had not invented the telephone by February 14, 1876, the same could be said of Bell. The letter surely carried a lot of weight at the trial; it torpedoed the case for the defense.

The Dowd case was only the most famous and not even the most protracted of the Bell suits. Without relating the morbid details, I probably shouldn't forget Daniel Drawbaugh, a blacksmith (among his other occupations) from the backwoods of Pennsylvania. Hearing in 1880 that Drawbaugh had invented a telephone, some investors applied for patents in his name, formed the People's Telephone Company, and were promptly sued by the Bell Company for infringement. What distinguishes this business from the others is the derisive attitude of the Bell camp toward Drawbaugh, Bruce simply dismissing him as "a charlatan." "When asked in court to explain the mental and experimental processes leading to these inventions," both Bruce and Mackay relate, "Drawbaugh replied, 'I don't remember how I came by it. I had been experimenting in that direction. I don't remember of getting at it by accident either. I don't remember of any one telling me of it. I don't suppose any one told me.'"

Not the best witness in his own defense. James Mackay goes on to tell us that the Bell lawyers "demolished" Drawbaugh. If so, it surely was history's most protracted demolition job. The case dragged on for a full eight years; unmentioned by Bruce and Mackay, the charlatan managed to stave off destruction till he was cornered at the Supreme Court. The final decision in Bell's favor was split four to three. Given that the Supreme Court sees fit to appoint presidents, it is not entirely obvious where the charlatans were sitting, but the minority opinion held that "overwhelming" evidence proved that as early as 1871 Drawbaugh had produced "articulate speech at a distance by means ... of a process substantially the same as that claimed in Mr. Bell's patent," and even the majority agreed that over a hundred witnesses had testified to this, though those folks couldn't explain how it was done. The majority agreed with Bell that the witnesses must have been mistaken because the instruments couldn't have done what the charlatan claimed.

The Dolbear case was the other famous one. Dolbear was an acquaintance of Bell's and a professor at Tufts College. In 1879 Dolbear invented a "static" telephone; it used not a vibrating membrane but rather two metal plates that formed an electrical component known as a capacitor.* Bruce calls this "essentially a Bell telephone with a condenser [capacitor] rather than a closed circuit." Coe calls it a fundamentally new way of constructing a microphone, with which I would tend to agree. Inevitably, the Bell Company sued Dolbear, and the case also reached the Supreme Court. In a decision still cited today in patent cases pitting broad ideas versus specific devices, the Court upheld Bell on the grounds that his patent covered any method of transmitting speech by undulatory current. Dolbear, like Gray growing increasingly bitter over the years, also claimed to have an invented a telephone prior to Bell—and students hung a plaque at Tufts to prove it, or at least to prove they believed it.

As you see, the telephone tango only began with the device's invention; it became more tangled with each passing year. The best was yet to come. By almost any reasonable account, Bell's transmitter—which had no amplifier— was simply not suited for wide networking, and this gave Thomas Edison the opening he needed. Tom tried placing a little button of carbon on the transmitter membrane, which would in turn press against a second carbon button. These small buttons, fashioned from soot collected from kerosene

* Speaking into the telephone varied the capacitance and created the undulating current.

lanterns, vastly improved the telephone's performance, and all this soot may well have given Tom's associate Charles Batchelor the idea of trying to make it into carbon filaments, which resulted in a well-known lamp.* Edison's patent on the buttons put the Bell Company in a tight position because Watson couldn't come up with anything better. Luckily for Bell, at just the right moment he ran across Emile Berliner (1859–1929), who in 1877 had filed a caveat for a transmitter thirteen comfortable days before Edison. Bell hired Berliner in exchange for the rights to his invention, and that act may well have saved the Bell Company from utter and immediate extinction.

Yet in a twist conspicuously absent from Bell biographies, the Patent Office failed to award the Berliner patent until 1891. That's fourteen years from the date of filing. In those days, and for a long time afterward, there was no little suspicion that the Bell Company itself had held up the patent to be able to fend off Edison and the other poachers for that much longer, extending the lifetime of the patent to a full thirty-five years. When it was finally awarded, Bell stock soared, and public outrage became so great that the government brought another suit against the Bell Company, one that also reached the Supreme Court. But the government failed to prove fraud and Bell won again. By one computation, the Bell Company had avoided paying $15 million in royalties to Berliner. Back then $15 million meant something. And there's a certain irony in that Berliner's transmitter wasn't more than a modified version of Reis' original: a sound box covered with a metal diaphragm that vibrated against an adjustable metal screw.

But if Edison had come up with a better transmitter than Bell, Bell could claim a right symmetric revenge. You'll recall that Edison invented the phonograph in 1877, but with tinfoil as the recording surface, it didn't work very well, tinfoil having a tendency to rip to shreds. It was actually Charles Tainter, Chichester Bell (Graham's cousin), and Alexander Graham Bell himself who at Bell's new Volta Laboratory in Washington developed the wax cylinder. Edison was forced to eat crow, or maybe paraffin, and license the invention for the production of his recordings.

In looking over the whole mess, one is struck by the futility of leaving technical decisions in the hands of judges. It makes about as much sense as leaving moral decisions in the hands of scientists. Many of the questions

---

* Once again the coherer principle was at work.

disputed a century and more ago—such as the nature of the Reis tele-phone—revolved around highly technical nuances that most engineers and physicists today would be hard pressed to explain. This issue is obvi-ously still with us.

Yet at the same time one is pitying the judges who had to sit through all this, one can't help feel that the real culprit is American patent law. I don't know to what extent American regulations deviate from interna-tional norms, but Bell's application for an Austrian patent was rejected because Austrian law forbade patenting principles as opposed to devices. In granting Bell a patent for what amounted to the transmission of speech by electricity, and upholding it in myriad court decisions, the government created a monopoly that existed for an entire century. People will have different opinions on whether that monopoly was the best thing that ever happened to America or whether it blocked progress for decades. Recently companies have attempted to patent basic search routines of the sort found in any computer program, as well as the entire field of robotics. When hearing such grandiose claims, it's hard not to become alarmed—and convinced that some adjustments to the law must be made.

The other moral of the story is that history is mighty pliable, espe-cially in the hands of portrait artists—so pliable you end up convinced not so much that Napoléon was right about history being a fable agreed upon, but that history has dissolved in the hands of historians. Modern theoretical physicists are regularly accused of theorizing in a vacuum devoid of facts. Not to be outdone, historians seem to have divorced his-tory from the past, a theme to be broadcast once more by radio.

Who invented the telephone? I've said enough. Yet one can't help but feel that there is truth in the sad epitaph Elisha Gray scrawled on a scrap of paper found among his effects after his death:

> The history of the telephone will never be fully written. It is partly hid-den away in 20 or 30 thousand pages of testimony and partly lying on the hearts and consciences of a few people whose lips are Sealed.—Some in death and others with a golden clasp whose grip is even tighter.

14 / A Babble of Incoherence: The Wireless Telegraph, a.k.a. Radio

On December 12, 1901, or perhaps a few days later, the world was astounded to learn that the *dot-dot-dot* of the Morse code for the letter *S* had been transmitted by radio across the Atlantic Ocean from England and received early that afternoon in Newfoundland by Guglielmo Marconi and his assistants. That much is true.

What is also true is that, with the possible exception of the telephone, no other exhibit in the Contemporary Panopticon's Domain of Technology raises blood pressure so high as the invention of radio. One has only to spend a few moments surfing the Web to understand that *bipolar* has passed beyond designating the nation's mental disorder of choice to include the state of partisan bloodbaths over the radio's origins: Tesla vs. Marconi, Popov vs. Marconi, Marconi vs. Bose, Rutherford vs. Marconi . . . Even disputes over the television are less passionate, doubtlessly because the story is so complicated that only the editors of *Time* magazine claim to know who created it. Yet when it comes to the apparently more straightforward case of the radio . . . In a letters column to the *New Scientist* not too long ago, readers advanced each of the names just mentioned as the true "father of radio." The amusing thing was, not one of the letter writers proposed more than a single candidate.

The truth is, the early history of the radio is so murky that historians are only now beginning to sort it out.

On Friday evening, June 1, 1894, Professor Oliver Lodge, who was briefly sighted earlier in his championing of J. J. Thomson and the electron, delivered a lecture-demonstration at the Royal Institution. For his talk, Lodge had placed a spark gap transmitter in the library and a receiver in the lecture theater, about forty meters away, with three rooms and a staircase in between. To the delight of the audience he was able to receive the electromagnetic signals produced by the transmitter, signals that today we would try to eliminate as static. Lodge had not been the first person to imagine the possibilities of radio waves for telegraphy; neither did he attempt to transmit any "intelligence" with his device. But it was the first public demonstration of a radio transmitter, and the lecture, reprinted in the influential journal *The Electrician,* became widely read, both in England and abroad. A few months later, on August 14, Lodge repeated the demonstration at Oxford for the annual meeting of the British Association for the Advancement of Science.

According to Lodge (nearly forty years later), at this demonstration he transmitted Morse code. Memory is the dung fuel of priority disputes, and in this instance a number of memories quite remarkably improved, but the prevailing opinion (in England at least) is that Lodge actually did what he said. Indisputable, and more important for all that went after, is that at these lectures the professor from Liverpool demonstrated several devices that might be used to detect radio waves, including one he had dubbed a "coherer."

Coherer is a word that means almost nothing to anybody, which is as it should be, since it is virtually meaningless. When Hertz had originally detected radio waves, he did so merely by observing sparks that jumped between two wires. A spark gap receiver is not particularly sensitive, and so over the next years scientists expended considerable effort devising better detectors. The result was the "coherer," which was little more than a glass tube filled with small particles, typically iron filings. When a radio signal came through, the particles tended to stick together; this action dramatically lowered the electrical resistance of the circuit, allowing a current from a suitably placed battery to flow. The current, in turn, could either be measured by a meter or cause a click in a telephone earpiece.

The crude microphones employed by David Hughes in his pre-Hertzian detection of radio waves (chapter 6) were essentially coherers, since they depended on loose contacts being jostled together by the passage of a radio wave, and the similarity to Reis' telephone also can't be missed.* The basic principle, though, seems to have been discovered by the Swede P. S. Monk as early as 1835, and in 1884 an Italian physicist, Temistocle Calzecchi Onesti, invented a coherer to which he attached a telephone. Such claims, of course, perturb supporters of Edouard Branly, who is usually credited with the invention of the thing in 1890 but who called it a "radio conductor." "Molecular receivers" and the German *Fritter* were also proposed, all of which confirm that *coherer* means nothing because there is no reason it should mean anything. Lodge's "coherer" referred to the tendency of the filings to stick together, and the term stuck too. Unfortunately, getting the filings unstuck once they cohered was the major diffi-

---

* Other receivers along similar lines employed frog legs and cats' brains. See Vivian J. Phillips, *Early Radio Wave Detectors* (London: Peter Peregrinus, 1980).

culty, and so much work went into designing coherers that included elaborate alarm-clock-like clappers as part of the circuit to decohere them.

Although all this seems unbearably primitive from today's standpoint, the coherer transformed a laboratory curiosity into a commercial possibility, and to talk about early radios without discussing coherers is like talking about televisions without mentioning television tubes. The fighting shortly to commence is largely over these small glass tubes filled with powder, which, for reasons no one comprehended, stuck together when struck by radio waves.

When Lodge gave his lecture at the Royal Institution, Guglielmo Marconi, born in 1874 of an Italian father, Giuseppe, and an Irish mother, Annie, had just turned twenty. Biographers invariably mention that Marconi's great idea came to him during the summer of 1894, when he read an obituary of Hertz, who had died in January. This particular obituary, however, appeared in the scientific journal *Il Nuovo Cimento,* accompanied by a discussion of Hertz's work, and was authored by the same Augusto Righi whom Italians advance as discoverer of the photoelectric effect. Righi had also researched Hertzian waves and had built upon the work of Lodge.*

As it turns out, Righi and the affluent Marconis were neighbors, and several years earlier Annie had arranged for Guglielmo to work informally in Righi's lab at the University of Bologna. Upon return from vacation Guglielmo announced to Righi his idea that Hertzian waves might be used to transmit telegraphy. Marconi later downplayed and even outrightly denied Righi's influence, but Righi himself had another opinion. In 1897 he wrote to Lodge: "I shall be very curious to know about [Marconi's] apparatus, but I suspect it resembles what he rigged up here with my oscillator and your coherer." Righi is correct in that Marconi's first transmitter was almost a direct copy of his own. In this light it is not too surprising that Righi discouraged the young man from pursuing his vision; at least discouragement is the version from the Marconi camp. Annie, though, had complete faith, and Guglielmo promptly set about performing experiments in the attic.

Marconi was underequipped for the adventure. He had been something of a solitary child, not popular with the gang. Due to his mother, he spoke English with native fluency, and when he was thirteen he had enrolled at the Leghorn Technical Institute, where he developed a single-minded pursuit of studies in physics and chemistry. With Annie's backing

* Although I find no obituary, the article definitely exists.

and his father's disapproval, Guglielmo even took private lessons in physics at the local lyceum, but it proved insufficient to get him into the University of Bologna, at which point Annie made her case to Righi. So Marconi had a technical background that eclipsed Morse's, but it was far from brilliant.

He was, however, persistent. Within months Marconi was sending radio signals across the attic room and making coherers—certainly improving Righi's—and by the summer of 1895 he was transmitting up to "a mile or thereabouts." During that time he also made two important discoveries. The first was that he could receive signals at greater distances if he held high a metal plate attached to one end of the coherer and buried in the ground a second plate attached to the other end of the coherer. The second was that a local hill provided no obstacle to radio transmission.

At this stage Marconi needed financial backing to proceed. Annie, who was from the Jameson whiskey family and had many contacts in England, got to work, and soon the two of them moved to London with a demonstration apparatus and letters of introduction. From that moment on, Marconi's progress was unimpeded. He immediately found backers, in particular the chief engineer to the post office, William Preece; he patented his device (on June 2, 1896); he started the Marconi Company; he made public demonstrations on Salisbury Plain. The British navy tested his wireless in combat maneuvers and adopted it. He continued to increase the range of his transmitters. The newspapers reported everything. Marconi became world-famous.

Oliver Lodge, for one, was not impressed. He wrote to the *Times* reminding readers of his demonstrations of 1894; he mentioned Marconi's "instructions" from Righi; he lauded Marconi for his hard work in turning the wireless into a commercial success; he criticized the press for their hyperbole in Marconi's favor. "The only 'important discovery' about the matter was made in 1888 by Heinrich Hertz," said Lodge, "and on that is based the emitter of the waves."

This was not mere sniping by a miffed outsider. Everyone in European scientific circles knew of Lodge's work, and when the details of Marconi's patent were finally made public in 1897, *The Electrician* jumped in with a blistering editorial:

> Dr. Lodge published enough three years ago to enable the most simple minded "practician" to compound a system of practical telegraphy without deviating a single hair's-breadth from Lodge's methods. . . . It is reputed

to be easy enough for a clever lawyer to drive a coach and four through an Act of Parliament. If this patent be upheld in the courts of law it will be seen that it is equally easy for an eminent patent-counsel to compile a valid patent from the publicly described and exhibited products of another man's brain. No longer is it necessary to devise even so much as "a novel combination of old instrumentalities," and the saying "*ex nihilo nihil fit*" [nothing comes from nothing] evidently was not intended to apply to English patents at the end of the nineteenth century.

No evidence exists that Marconi had gotten his first ideas directly from Lodge. But directly or indirectly, it was a debt, and in that sense *The Electrician*'s position is defensible. Lodge himself by that time had been spurred to action. A scientist who published all his work in the open literature, he had not initially intended to patent his discoveries (and even admitted that it had not occurred to him), but immediately after the August 1894 demonstration his colleague Alexander Muirhead began urging him to do so. Tradition has it that Lodge resisted, but this may not be so.* One thing is certain: For all his public politeness, Lodge became increasingly angry with the publicity that Preece—a mediocre scientist, a great showman, and a personal enemy—was handing to Marconi in theatrical and extremely popular lectures. At one point Preece even began to claim that Marconi had discovered a completely new form of ray, an assertion Marconi repeated in print! Such shenanigans were designed not only to create a legend (which evidently succeeded) but also to ensure that Marconi's invention was patentable.

British law gave precedence to the first applicant, not to the inventor, and strong evidence exists that Marconi's advisors recommended that he patent *everything*—even the use of Hertz waves themselves—then defend. Indeed, Lodge would later criticize Marconi for "his tendency to attempt a claim at everything." But Lodge did not merely protest; he patented. In May 1897, while the details of Marconi's system had yet to be made public, Lodge patented all aspects of his system he was certain Marconi had not duplicated. At its basis was the concept of syntony, a beautiful word that should be resurrected.

Spark gap transmitters essentially broadcast static, which splashes across all frequencies, and any two transmitters in the same vicinity interfere with

---

* There now seems to be evidence uncovered by Patrick Muirhead (Alexander's great-nephew) that Lodge and Alexander Muirhead registered a patent for electrical telegraphy on April 23, 1895, which was subsequently abandoned either because it was withdrawn or blocked (perhaps because, according to British patent law, one could not patent a device that had been publicly exhibited). At the present time it is not known which. I thank Peter Rowlands for this information.

each other, rendering communication virtually unintelligible. Furthermore, the very nature of the signal makes eavesdropping not only inevitable but unavoidable, and secrecy impossible. With the rapid increase in the use of radio, it became essential to find some way of syntonizing the transmitter and receiver to the same frequency in order to screen out interference. This is what Lodge did with his master 1897 patent.

In March 1900 Marconi himself took out a patent for syntony, which he termed "tuning." Patent 7,777, the famous "four sevens," was destined to become the subject of more litigations than virtually any patent in radio history. And for good reason: It was a blatant infringement on Lodge's. Marconi's version, it can't be denied, was a significant improvement over his competitor's, but it used all the features that Lodge's patent covered. It is fruitless to argue that the Marconi Company was then unaware of Lodge's published work. Rather, they apparently guessed that Lodge wouldn't put up a fight, and for eleven years the Marconi Company paid no royalties to Lodge and Muirhead, who had meanwhile formed the Lodge-Muirhead Syndicate. Why Lodge allowed himself to be steamrolled remains unclear, but he gradually collected a war chest and eventually went to battle. In 1911 the courts extended his patent for a further seven years, which made Marconi's legal position untenable; the Marconi Company bought out the Lodge-Muirhead Syndicate and tuning patent. Lodge was retained as a consultant to the Marconi Company, but according to his own account he was never consulted about anything. Biographies of Marconi, not too different in this respect from those of Bell, tend to summarize the entire affair in one sentence: The Marconi Company acquired the Lodge patents in 1911.

By this time a pattern is recognizable. Marconi, with his limited technical background, was interested only in making radio a commercial success. He relied on others for technical help and borrowed whatever intellectual property he needed to achieve his aims. This amounted to a certain unscrupulousness that manifested itself time and time again in dealing with his competitors, as will become clear in another small episode that tends to be overlooked by biographers.

The Italian navy coherer scandal has much in common with the Schleswig-Holstein question, whose legendary complications prompted Lord Palmerston to remark that only three people had ever understood it: Prince Albert, who was dead; a professor who had gone mad; and Palmerston himself, who had forgotten it. The affair began shortly after Marconi's return from Newfoundland and the 1901 transatlantic experiment that so astonished the world. You must understand that at the time scientists had been shocked to learn that radio waves could be transmitted beyond the horizon. But no one besides Marconi believed that a radio signal could be transmitted across an ocean, and the Marconi legend largely rests on the epic experiment that proved everybody wrong. For the test his company had erected a high-power spark gap transmitter in Poldhu, Cornwall. No one has denied that Ambrose Fleming, later inventor of the vacuum tube and advisor to the firm, designed the transmitter, but once home Marconi faced accusations that he had stolen the design for the receiver. The question has always been—and remains—from whom.

In the summer of 1901 Marconi's childhood friend, Marchese Luigi Solari, had come to London to borrow some of the synotised apparatus for testing by the Italian navy, in which Solari served as a lieutenant. In return for the loan Solari presented Marconi with a newly invented coherer, which, unlike the usual ones filled with iron filings, consisted of a tube containing a drop of mercury between two metal plugs. (An advantage of this arrangement was that no filings remained stuck together and so the device automatically decohered.) When Marconi inquired as to the origins of the device, Solari told him only that it had been developed by several members of the navy and so should be known as the Italian navy coherer. Thus it became. With the blessings of the navy, Marconi promptly applied for a patent in his own name. The thing might have been forgotten except that this coherer became the essential component in the receiver that fished the transatlantic *S* out of the ether on that immortal December day.

Winds of war began to gust in 1902 when Professor Angelo Banti published an article in his journal *L'Elettricista* claiming that the mercury coherer had been invented by a lowly navy signalman named Paolo Castelli. The article caused such a furor that Banti was forced to provide more evidence—which he did in the form of reports direct from the Italian navy. Later, as Banti obtained more documents, it began to appear that Solari and Castelli had developed their coherers independently. Banti penned a "recantation," which ended by saying that he "had not published any false thing." Instead "there has been raised a question of soups and sops [apparently, idiomatic Italian for tempest in a teacup] for the purpose of concealing some maneuvers to which we are strangers. And this suggests that . . ."

This is exactly how Banti's piece ends, and the implication of the ellipsis is clear. Maybe the invention was simultaneous. On the other hand . . .

In England, Silvanus Thompson, as ardent a supporter of Lodge as he had been of Philipp Reis, was attacking Marconi in the pages of the British *Saturday Review*. In a fifteen-round slugfest, Thompson accused Marconi of pirating Lodge's coherer and that the antenna and the connection of the coherer to earth was "the only real point of novelty . . . All else is simply detail or surplusage."

Marconi countered that Lodge's patent applications were made after his own and that the antenna was the feature that "made the whole difference between workable and unworkable." Lodge's attempts to make a practical wireless system were "entirely fruitless." Take that. But Thompson was back with a sharp rebuttal to the left, detailing why Marconi's patent claims were balderdash in light of Lodge's, and now he demanded point-blank whether Marconi had ever heard of Castelli. Marconi is buckling. No, yes. Is he going to punch? No, Marconi caves, replying only that he would "decline to take in hand the tediously long task of enlightening the Professor." The crowd is going wild.

Thompson's opponent totally evaded the Castelli issue and from that round on remained silent, boxing only through proxies, who included Solari and even a member of Parliament. Solari, in a letter to the *Times,* claimed that the coherer had been "devised in the stations under my own immediate supervision," but he did not want to patent it because "I have always been dissuaded from such a recourse by recollection that the idea of the employment of mercury had been suggested by something which I had read in some English publication which I found myself unable to trace." Marconi, in a Royal Institution lecture of June 1902, revealed that to clear up the matter he had written to the minister of the marines, who replied that the coherer "must be considered the work of various individuals in the Royal Navy, not of one."

This, evidently, was taken by Thompson as a tacit acknowledgment of Castelli's work, for he now jumped in again with a blistering letter to the *Times,* pointing out the inherent contradictions in the Italians' position. "What all this cloud of dust is designed to conceal I do not know. The Lieutenant says that the petty-officer did not invent the tube, and gives us his *ipse dixit* that he invented it himself. That captain says on the other hand that the petty-officer 'invented' it, 'constructed' it and 'proposed' its use. . . . If the Lieutenant is right, the Captain is wrong, and if the Captain is right, the Lieutenant is wrong."

Thompson scored a KO: Within a month Marconi "converted" his patent claim into "an application for an invention communicated to him from abroad by the Marquis Luigi Solari of Italy." The capitulation struck—and continues to strike—many as de facto proof that a cover-up was in the works. Yet this was far from the end of the affair. A year later E. Guarini published a letter pointing out that the real inventor of the mercury coherer was neither Solari nor Castelli but Professor Thomas Tommasina of Geneva.

Tommasina, also not mentioned in certain biographies, had indeed published a paper on May 1, 1899, in the *Comptes Rendus* of the French Academy of Sciences, in which there is one line referring to a mercury-drop coherer. Tommasina himself began to insist on his priority and presented further documentary evidence, and even some specimens of coherers to support his claim. Forty years later Solari would claim that, truth be told, the Italian navy coherer was based on Tommasina's work.

Solari's continued evasiveness inevitably suggests that unsavory practices were afoot, probably theft of Castelli's design, but because no one was able to prove anything the matter eventually languished, and Marconi shared the 1909 Nobel prize for physics with his German competitor Karl Braun (founder of Telefunken and also little mentioned in certain biographies). Then deathly silence for nearly a century, until quite recently, when engineer and historian Probir Bondyopadhyay threw another monkey wrench into the chassis. His claims, made in a detailed article that appeared in the *Proceedings of the IEEE,* were picked up by Indian newspapers as well as *Science* magazine and caused a microscopic international furor. Specifically, Bondyopadhyay presented "incontrovertible evidence" that the notorious coherer was invented by the Indian Jagadish Chandra Bose and that Solari had simply stolen it.

In every respect a unique character, Sir Jagadish Chandra Bose is regarded by Indians as not only India's first physicist but also the country's first plant physiologist. And the inventor of radio. Born in Bengal in 1858, Bose first studied at the University of Calcutta, from which he graduated with a bachelor of arts degree in 1880. He then traveled to London to study medicine, but his medical studies were quickly derailed, and in 1882 he entered Christ's College, Cambridge, to read natural sciences. There he took courses in physics from Lord Rayleigh himself and also studied chemistry and botany. Returning to Calcutta in 1885, Bose was appointed the first Indian professor of physics at Presidency College, where for the next eight years he did little research. Suddenly, at the age of thirty-six in 1894, he began a series of experiments in radio that was to last six years and end nearly as abruptly.

Around 1900 Bose's attention drifted from physics toward plant physiology. He developed a thesis that no firm distinction existed between the living and nonliving and that organic and inorganic matter responded similarly to external stimuli. The idea led its enunciator into controversial territory, in which he subjected plants to electrical and thermal impulses and announced that the ascent of sap in trees was due to protoplasmic rhythmic pulsations. Ultimately his fringe speculations prompted colleagues to denounce Bose's work as "a danger and a menace to sound science." His reputation never fully recovered, and he died in 1937.

Regardless, Bose's early work on radio rivaled and often anticipated those of his peers. Having read Lodge's famous lecture, Bose began to duplicate Hertz's experiments showing that radio waves could be reflected, refracted, and polarized, except that he managed to do this in what we would call the microwave spectrum—wavelengths ten times shorter than those of his contemporaries. He made his first public demonstration of radio transmission in 1895, using the waves to detonate gunpowder and ring a bell. Braun, in 1874, had discovered that certain substances, in particular the mineral galena, conducted electricity in one direction but not in another. This was the discovery of the semiconductor diode (and the explanation of Braun's Nobel prize). Bose found that galena could be used to detect both radio and optical waves, and on a Friday evening in May 1901 he demonstrated before the Royal Institution an "artificial retina" that consisted of such a galena detector; in 1904 he patented a device employing galena to detect radio signals. It was the first crystal radio, beloved of all boy scientists now over forty-five. In this field Bose was, it has been said, "at least sixty years ahead of his time."

And in 1899 Bose invented a mercury coherer, an account of which (as with many of his discoveries) was communicated to the Royal Society by Lord Rayleigh on April 27, five days before Tommasina's paper was published in the *Comptes Rendus*.

Having studied Bondyopadhyay's article, I have not perceived the smoking gun he insists points to Bose as the originator of the Italian navy coherer. The "incontrovertible evidence" is mainly Bose's *Philosophical Transactions* paper, which Bondyopadhyay asserts must be the "English publication" Solari found himself so strangely unable to trace. To be sure, Bose describes a mercury coherer, but the description amounts to one paragraph in a seven-page paper. Bondyopadhyay claims the differences between Bose's coherer and the one used by Marconi are trivial, but Bose presents no specifications and no diagram. As Subrata Dasgupta points out in his biography of Bose, the differences between the Castelli and Tommasina coherers, on one hand, and Marconi's, on the other, are probably no greater than those between Marconi's and Bose's.

It cannot be denied that Marconi knew of Bose. Famous back-to-back interviews with the two inventors were published in the March 1897 issue of *McClure's Magazine;* it was in this interview that Marconi claimed he had discovered a new kind of ray. Also in Bondyopadhyay's favor is that it is difficult to see what other "English publication" Solari could have been referring to. Shortly after his first letter to the *Times,* Solari did claim to remember it: John Fahie's *History of Wireless Telegraphy.* But the only place in Fahie's book where mercury coherers are discussed is the reprint of Marconi's own 1896 patent application! They are not exactly globule coherers (the mercury is mixed in), but Solari's pleas begin to seem a bit circular.

Nevertheless, Marconi clearly had mercury on the brain prior to 1899. According to Dasgupta, several other Italians also experimented with mercury globule coherers in 1900. Neither was Bose the first to attach a telephone to the receiver, another piece of evidence Bondyopadhyay enlists to demonstrate that Solari pilfered the Indian's invention. The unfortunate David Edward Hughes employed exactly this technique even before Hertz and Onesti did the same. Finally, Bondyopadhyay's logic seems curious. He attempts to prove the Italians were liars, yet adduces such statements as "Marconi clearly stated that the device he used to receive the first transatlantic wireless signal was not invented by Tommasina. Therefore, subsequent claims attributed to Tommasina by others or as claimed by Tommasina himself are totally fraudulent." One is reminded of the Cretan paradox. And so as the sun sets in the west, Solari was almost certainly involved in grand theft, but were I sitting on a jury, I would probably have to acquit on the grounds of habeas corpus—the victim has not been found.

None of this matters, of course, because the real inventor of the radio was Aleksandr Popov. In that legendary Russian portrait gallery of inventors invented by Ivanov, Popov's portrait is the grandest. For decades the Communists proclaimed him the true father of radio and elevated him into a national hero. Today, a decade after the collapse of the Soviet Union, you can still see his first apparatus at the Moscow Polytechnical Institute and you will still be told he was the inventor of radio. Popov, born in 1859, studied at the University of St. Petersburg and became a faculty member at the Russian navy's torpedo school on Kronstadt, which was considered a choice position. He had also read Lodge's famous lecture and set about improving the Englishman's coherer. In addition, Popov added to the circuit a relay with a clapper that tapped loose the iron filings in the coherer at the same time it rang a bell. On May 7, 1895, he demonstrated to the Physico-Chemical Society that this apparatus was capable of detecting distant thunderstorms. This was nearly a year before Marconi's first demonstrations in London, hence the Russian claim that Popov was "first."

Unfortunately, Popov never patented his detector. In terms of proof of concept, though, the demonstration did take place; Popov also published clear descriptions of trials and device (with diagrams). And although Marconi's first patented receiver seems almost identical to Popov's, no one, least of all Popov, ever claimed that Marconi swiped it. Indeed, after his initial demonstration Popov got distracted by X-ray research, and the news a year later that Marconi had succeeded in transmitting signals over kilometers "shocked him out of his torpor." Popov returned to radio work, soon marketing his equipment through Eugène Ducretet in Paris. Still, Popov never held a grudge against Marconi for cornering the spotlight. Rather, he freely acknowledged that "it is beyond all question that the first practical results in wireless telegraphy over considerable distances have been achieved by Marconi."

Popov's reputation in the West, though, was hoisted on the Communist petard. Ivanov painted his portrait so brilliantly that Western authorities became convinced it was a forgery. Specifically, in 1925, on the thirtieth anniversary of Popov's first demonstation, a Soviet official, Victor Gabel, wrote to *Wireless World* magazine claiming that Popov had not only received a signal on May 7, 1895, but transmitted one as well, the words "Heinrich Hertz." Upon a skeptical inquiry from the *Wireless World* editor, Gabel produced three apparently credible eyewitnesses, but the date they remembered was March 24, 1896. To this day no documentary evidence has been found. The eyewitnesses claimed that Popov himself asked that nothing be recorded because he worked for a military organization. Nevertheless, in 1945 the Soviet government declared Popov the inventor of radio, a claim that has yet to find peace.

But Popov's existence and his contributions cannot be denied. In 1899 he set up wireless communication with the battleship *Aprashkin,* which had run aground on Hogland Island in the Gulf of Finland. Communications lasted several months until the ship was freed, during which time wireless telegraphy saved fifty Finnish fishermen—an ice floe had broken loose with the men on it, and via wireless an icebreaker en route to the *Aprashkin* was diverted to rescue them. The story seems to be true; certainly officials at the Polytechnical Museum believe it. And just as combat maneuvers convinced the British navy to adopt the wireless, the *Aprashkin* episode convinced the Russian navy to do the same. Popov oversaw the equipment's installation.

Unfortunately, Popov's scientific work was soon to be curtailed. In declining health, possibly worsened by exposure to X rays, he died in January 1906. By that time he was a broken man, devastated by a tsarist order to repress the student uprisings at the St. Petersburg Electrochemical Institute, where he had just been appointed director.

A Russian story would hardly be a Russian story if it had a happy ending, would it? We are saved from too much morbid contemplation because there is not the slightest doubt that the radio was invented by Nikola Tesla. The *New York Times* declared it so. This will not come as any surprise to Teslites the world over, who have been known to claim that Tesla's brilliance so eclipsed that of ordinary humans that he could only be an extraterrestrial, and who are perpetually determined to rescue him from obscurity despite the fact that he has been the subject of numerous biographies, plays, and documentaries; that at least three organizations exist to promote his name; and that he has been honored with a U.S. postage stamp, not to mention the international unit of magnetism. All this apparently does not suffice, and Teslites claim for him, in part, the invention of the laser, the computer, the atomic bomb, the electron microscope, robotics, the Strategic Defense Initiative, microwaves, and nuclear fusion. It goes without saying that everything in Tesla's wake has been a mere mop-up operation. Of course he invented the radio.

Nikola Tesla, a Serb born in Croatia in 1856, was a great inventor. No one has ever denied that he created the multiphase motor and the system of AC power transmission in universal use today. That story, at the center of the infamous "current wars" waged between Edison and Westinghouse at the close of the nineteenth century, has been touched on in chapter 12. The Teslites' claim that Tesla invented radio rests on three basic contentions: that in 1892 and 1893 Tesla "proposed the essential elements" of radio, that U.S. courts rejected Marconi's patent application in 1900 in favor of Tesla's earlier patent of 1897, and that in a famous 1943 decision the U.S. Supreme Court overturned Marconi's patents in favor of Tesla's.

As to the first contention, in 1893 Tesla, on a European lecture tour, consulted with Lodge and Crookes, with whom he shared an interest in psychic matters. Afterward he wrote up his proposals and published them in the *Journal of the Franklin Institute*. Some of his ideas were very close to those of Lodge's friend Fitzgerald, of the Lorentz contraction, but in any case Tesla built nothing at the time. He merely proposed.

As for the second contention, it was presumably the U.S. courts' rejection of Marconi's 1900 patent in favor of Tesla that prompted *New York Times* reporter William Broad to join the ranks of the Teslites and announce, "It was Nikola Tesla, not Marconi, who invented the first radio."

But Marconi's 1900 patent was not for the first radio, it was for his tuning system, which infringed on the earlier one of Lodge.

Indeed, contrary to what Teslites invariably claim, the Supreme Court in its landmark decision of 1943 never asserted that Tesla invented radio.* Rather, it threw out Marconi's tuning patents in light of the earlier ones of Tesla, Lodge, and inventor John S. Stone. In fact, the Court seemed to feel Stone's work was the most important, and who's ever heard of him? He must have been Tesla in disguise.

What are we to make of all this? Ask Rutherford. In early 1894, quite independently of the others, Ernest Rutherford in New Zealand began experimenting with radio, using a magnetic detector rather than a coherer. A year later he came to Cambridge with his apparatus, continued to demonstrate it, and for a time held the distance transmission record, one and a quarter miles. One can't help but feel the history of radio might have been written much differently if Rutherford hadn't gotten distracted by other things, such as discovering the nucleus of the atom. One might also emphasize that all these systems were designed solely for transmitting Morse code. The Canadian Reginald Fessenden made the first voice broadcast in 1906.

I repeat, what are we to make of all this? Beyond the obvious, in reading accounts of early radio, as of the telephone, one gets an extreme sense of the malleability of history. Contrast this statement by an early, authorized Marconi biographer, "There may never be another genius to whom science will award the sole honor of a great discovery and historically link his name with the invention," with this more recent evaluation by Sir Bernard Lovell: "Marconi was the forerunner of the modern pseudo-scientists and technicians who make fortunes out of scientific discoveries because of their business acumen."

It is perhaps more than malleability. More than in any other Panopticon exhibit, the various accounts of the telephone's and radio's origins describe almost nonintersecting histories. What is striking about the biographies of Marconi in particular, as of Bell, is not so much what they do say but what they don't. If what I have written is known to aficionados of early radio, it is conspicuously absent from more mainstream accounts.

---

* According to the author Hugh Aitken, the reason the case dragged on to such an extraordinary date was that the U.S. government had used the tuning circuits during World War I without paying anybody any royalties. Presumably they weren't any more anxious than Marconi to do so.

Why? History, they say, is written by the victors. Although Marconi occasionally acknowledged his predecessors, such acknowledgments became ever more miserly, and by the time of his Nobel lecture in 1909, Righi is mentioned only in passing and Lodge in a footnote. By the time of the authorized biography, 1894 has been erased. The art of propaganda indeed lies not in sins of commission but in sins of omission. Though sometimes of commission, too: Ambrose Fleming, while a consultant to the Marconi Company, flatly denied that Lodge had transmitted telegraphy in August 1894, then eventually changed his mind, declaring, "It is, therefore, unquestionable that on the occasion of his Oxford lecture in September, 1894, Lodge exhibited electrowave telegraphy over a short distance." That he got the month wrong is not inspiring, but less so is that by his own admission he was not present at the event and was relying on information supplied by Lodge. On the other hand, Marconi's daughter, in her biography of her father, changed Fleming's "unquestionable" to "questionable," all of which shows brilliantly the oscillations of history.

It goes without saying that adventurers with business acumen are necessary to bring a technology from the laboratory stage to commercial practicality, and it is worth pondering the relative contributions of Henry and Lodge, and Morse and Marconi, in their respective dramas. It should also go without saying that the proper function of scientific and technological history is to illuminate the process of cross-fertilization and the development of ideas. If denizens of the Internet and electrical magazines would attempt to understand how the various pioneers of wireless telegraphy interacted and influenced one another, everyone would profit. The promotion of personal or national heroes is not history; it is boxing.

15 / Mind-Destroying Rays: Television

"All tales should be embellished," the physicist Freeman Dyson avows. Thus ten, perhaps fifteen years ago, he related the following account one fine summer day as we were rowing strenuously along Lake Carnegie in Princeton, New Jersey. A few fine summer days earlier, maybe a year, there had been a timid but persistent knock on the door of his house, which stands near the grounds of the Institute for Advanced Study. Dyson answered and found facing him on his doorstep a short, dark Russian, who asked shyly in heavily accented English whether this was the home of Vladimir Zworykin. No, Dyson replied, that was next door. The stranger explained in his broken English that he was writing a biography of Zworykin and wanted to look at the house-museum. So Dyson took him next door and they knocked. The residents, who were only renting for the summer, came to the door, and the would-be biographer explained that he was writing a biography of Vladimir Zworykin and was the house kept as a museum and, if so, would it be possible to visit. The couple had never heard of Zworykin, but since they seemed to be living in his house and since the Russian intended to write a book about whoever Zworykin might be, the guest was welcome to have a look. So he and Freeman went in. No sign of Zworykin.

Afterward the two of them drove across Route 1 to RCA's David Sarnoff Research Center. Again the Russian explained his mission and wondered if it would be possible to examine the Zworykin archives from the 1920s and 1930s, but the receptionist had never heard of Zworykin either. Neither had any of the lab personnel they encountered. Finally the vice president in charge of research and development told them that if any archives of this person had ever existed, they had been discarded decades ago. "Who was Zworykin?" the VP finally asked.

"He invented television," replied the Russian. "Here."

"I thought an Englisman invented television," a British friend of mine remarked when I recounted the story to him. Yes, and there was the Japanese fellow, and the Germans . . .

Why anyone would want credit for the invention of television is beyond me. Zworykin himself refused to watch it and must have been grateful for the "television paradox": Although it is arguably the most important invention of the twentieth century (certainly the most pervasive), its history is one of the least known to the general public. Yet this has not stopped enthusiasts from battling over who created it. "In the heated historical debate," wrote the editors of *Time* magazine of their profound deliberations over the most influential inventors of the twentieth century, "both TIME and the US Patent Office ended up giving [Philo Farnsworth] credit for the invention over his rival Vladimir Zworykin of RCA." *Time*'s egregious error was not in coming down on the side of the wrong man but that it came down on the side of a single man. If there is any major invention that was the product of work by dozens, even hundreds of individuals, it was television. As in any TV movie "based on real events," we can't recount all the twists and turns of the epic and will content ourselves only to highlight the major developments.

In 1880 Alexander Graham Bell deposited at the Smithsonian Institution a sealed cache of documents concerning an invention he termed the "photophone." Rumors immediately sprang up that the mysterious photophone was a method for seeing by telegraph. In one of the earliest and best examples of RASEP (result amplification through stimulated emission of publicity), the hearsay spurred a flood of rival inventors to disclose projects they already had in the works. A number of these involved "selenium cameras" or "selenium mosaics." Seven years earlier Willoughby Smith had discovered that the electrical resistance of the element selenium varied with exposure to light. It soon occurred to tinkerers the world over that selenium might provide the basis for the electrical transmission of images, and numerous proposals were advanced for doing so. Most involved an array, or mosaic, of selenium cells that would be exposed to light and connected by wire to some sort of detector. The detector would take the fluctuating electrical currents and retranslate them into an image by mechanical or electrochemical means.

Bell's shadowy photophone itself involved selenium, though when in August 1880 he revealed the contents of the sealed documents, they turned out to have nothing to do with seeing at a distance. At the core of his proposal was only a mirror that vibrated in response to speech. Due to the vibrations, a light beam bouncing off the mirror would vary in intensity, and this variation could be detected by a selenium cell at a distance, thus allowing the speech to be re-created. It may have had nothing to do with seeing at a distance, but Bell's "practical joke" inadvertently had a larger effect: It touched off the marathon race for television.

The problem with the selenium mosaic was that each element in the array had to be connected by a separate wire to the receiver, which could easily amount to a rat's nest of hundreds if not thousands of connections. An important conceptual advance took place before the end of 1880 when a Frenchman, Maurice LeBlanc, published an article in *La Lumière Electrique* suggesting that the selenium array could be replaced by a single cell if one scanned the image sequentially. (The method LeBlanc proposed involved two oscillating mirrors that swept across the illuminated image, but this is not so important.) Scanning each part of the image in turn would allow a signal to be transmitted serially via the selenium cell to a detector, allowing one to dispense with the multiple connections. LeBlanc also argued that because of persistence of vision, rapid sequential scanning would allow one to build up a likeness of the original image. This is a rather remarkable idea for 1880 (cinema had yet to be invented) and one that modern television relies on.

Within four years Paul Nipkow of Berlin showed how to make LeBlanc's idea practical and received one of the most—if not *the* most—important patents in television history. Nipkow proposed that the scanning be accomplished by a simple disk perforated with holes lying along the curve of a spiral. As the disk rotated, each hole in the spiral would pass in front of a different part of the image, allowing light to pass through to the selenium cell. At the receiving end, the varying current produced by the selenium cell controlled the intensity of a light source that was viewed through a second disk rotating in synchronization with the first. In this way the image could be reconstructed. Nipkow never actually built his device, but others did, and it became the basis for nearly all early television scanners. The disk, however, brought with it other problems. The rapid passage of the holes in front of the image diminished the amount of light that actually hit the selenium cell. This required not only more sensitive cells but cells with faster response times, and much effort went into developing them. One should not dismiss the Nipkow disk as a primitive contrivance. Mechanical televisions based on it persisted well into the 1930s, and even today advanced research television systems employing the disk are marketed because of its reliability and comparatively low cost.

Perhaps the next important advance took place in 1897, when the same Karl Braun who tends to get eliminated from Marconi biographies perfected the cathode ray tube, or CRT, which without much modification forms the picture tubes in modern televisions. Electrons are given off at one end of the tube, then deflected by electromagnets (or charged plates) so that they hit and produce a bright spot on a fluorescent screen positioned at the far end of the tube. In Braun's original design the electrons could be deflected in only one direction, but if they could be swept both horizontally and vertically, then one could build up a pattern of bright and dark spots—a picture.

The Braun tube, as it was called back then, soon figured in one of the most significant developments in the history of television, which proved to be not a device but a letter. In 1908 a Briton named Shelford Bidwell wrote to *Nature* magazine, reviewing some of the proposals for "distant electric vision." He concluded that to transmit even a two-inch-square picture would require ninety thousand selenium cells, with attendant lenses and wires; the receiving apparatus would occupy four thousand cubic feet, and the connecting cable would have a diameter of about ten inches, all of which could be had for a cost of £1.25 million.

Bidwell's letter brought a swift response from the eminent British engineer Alan Archibald Campbell Swinton (1863–1930), whose letter in the June 18 issue of *Nature* outlined almost all the important aspects of what would become electronic television. He pointed out the impracticality of effecting "160,000 synchronised operations per second by ordinary mechanical means," then added that "this part of the problem of obtaining distant electric vision can probably be solved by the employment of two beams of kathode rays (one at the transmitting end and one at the receiving station)." He then goes on to propose using a CRT in which the electrons can be deflected both horizontally and vertically. "Indeed," Swinton continued, "so far as the receiving apparatus is concerned, the moving kathode beam has only to be arranged to impinge on a sufficiently sensitive fluorescent screen, and given suitable variations in intensity, to obtain the desired results." Finally, he concluded prophetically, "The real difficulties lie in devising an efficient transmitter which, under the influence of light and shade, shall sufficiently vary the transmitted electric current so as to produce the necessary alterations in the intensity of the kathode beam of the receiver, and futher in making this transmitter sufficiently rapid in its action to repond to the 160,000 variations that are necessary as a minimum."

So accurate were Swinton's assessments that in Britain his letter is referred to as the "birth of electronic television." This is not much of an exaggeration, because although Swinton never built a working demonstration, once his vision became known worldwide after 1915 it guided the evolution of television to the system we know today. At the time, though, Swinton's readers may have been stunned rather than stimulated, because his suggestions went unanswered.

Perhaps it is better said that they had already been answered, in Russia. Nearly a decade before Swinton, Boris Rozing (sometimes spelled Rosing) had recognized the limits of mechanical television. One of the unsung pioneers of television, Rozing was born in 1869 in St. Petersburg and in 1887 entered the Department of Physics and Mathematics at the University of St. Petersburg. After graduation in 1893 he stayed on as assistant to the chair of physics, and in 1897 he was appointed director of the physics department at the Technological Institute of St. Petersburg, where he commenced work on distant electric vision.

For Rozing it was clear that the goal for electric vision must be "the elimination, insofar as possible of all inertial mechanisms from the electrical telescopes and replacing them by intertialess devices to the full extent of the significance of this word." In 1900 he made a formal report of his work to the International Congress of Electrotechnique, entitled "The Present Position of the Problem of Television." It was probably the first time the term *television* had ever been used.*

In 1902 Rozing began experimenting with a Braun tube for the receiver (the "television set"). In the system that emerged, the image was scanned by two rotating mirrors, converted to an electrical signal by a photoelectric cell, and the signal was sent on to the CRT; upon receipt of the signal, an appropriate burst of electrons was shot to the fluorescent screen. But to reconstruct the image properly, you need to tell the CRT exactly where to send the electrons. To accomplish this synchronization the mirrors also sent out a signal to the CRT, informing it how to deflect the electrons that corresponded to each part of the original object. By 1907 Rozing was able to patent his system, although at the time he had not successfully transmitted an image. He did so in 1911, transmitting an image of four luminous bands. This was the first time a picture had ever been transmitted and reconstructed by electronic means.

Rozing's system was of course not entirely electronic, relying on mirrors to scan the image, but it was the most advanced system of its day, incorporating, apart from the CRT, a fast photoelectric cell he had developed, and its importance was not far behind that of the Nipkow disk.

---

* Unless it was by Konstantin Persky at the same conference.

Rozing went on to improve his device, eventually transmitting up to 2,400 "picture elements," or "pixels." Unfortunately, Rozing encountered the same fate as many other Russians of his time. In 1931 he was falsely accused and convicted of anti-Stalinist activities and exiled to Archangel in the far north, where he died two years later of a cerebral hemorrhage. One of his experimental devices can still be seen in the Moscow Polytechnical Museum.

Perhaps Rozing's most important legacy, though, was not his television system but one of his pupils who worked with him at the Technological Institute: Vladimir Kosma Zworykin.

Zworykin, born in Murom in 1889, referred to being invited to work in Rozing's lab as "my most exciting moment . . . Through the door which he opened at that moment of invitation, I stepped with him into a new and challenging field, following a path that led ultimately to the iconoscope and to the birth of modern electronic television." By the time Zworykin stepped into Rozing's laboratory and onto the path toward electronic television (a path that would not go unchallenged), he had already been a student for several years at the Technological Institute, having graduated with honors in 1905 from the Realschule in Murom and having decided almost offhandedly to take up engineering at Petersburg. Whether due to the political upheavals of the times or for other reasons, Zworykin does not appear to have begun working with Rozing until early 1911; he was present at the May demonstration and remained an apprentice in the lab for over a year, until his graduation in 1912. Afterward Zworykin moved to Paris to study with the eminent physicist Paul Langevin, building equipment to study X-ray diffraction and, in his spare time, radio receivers. Realizing that his background in pure physics was weak, Zworykin decided in the summer of 1914 to remedy the deficit by studying in Berlin.

World War I intervened. Zworykin hightailed it back to Russia and spent a fairly adventurous few years as an army radio engineer. But after the war, the revolution. Zworykin, about to be arrested as an army officer who had not registered with the Bolsheviks, decided to flee the country. He made a daring escape eastward to Siberia, which during the civil war had refused to join Bolshevik Russia. At the city of Omsk he managed to join an Arctic expedition and set sail for Archangel, where, as it turned out, all the foreign embassies had relocated after the revolution. Zworykin convinced the various authorities to grant him papers, and on New Year's Day in 1919 he arrived in New York. But Zworykin didn't stay. The provisional government in Siberia asked him to return with his expertise and radio equipment. Rashly Zworykin agreed; he traveled back to Siberia, fulfilled his duties, and with some difficulty returned, once and for all, to the United States, in August 1919.

Zworykin soon got a job with Westinghouse in Pittsburgh, quit, moved to Kansas City, and then returned to Westinghouse. For the next six years Zworykin would inch toward the development of a fully electronic television, with both an electronic camera/transmitter and receiver/picture tube and no wires between. At the time, the development of each of these components was a tremendous technological challenge, and inventors the world over added incremental improvements. It was by no means obvious to those working in the field that electronic television would beat out mechanical systems, and numerous engineers pursued the latter. During much of the 1920s, in fact, mechanical systems were ahead.

As early as 1925 the Scotsman John Logie Baird, to the English probably the best-known television pioneer, had set up a television based on a Nipkow disk in Selfridge's department store in London for public display. Customers could peer through a narrow tube to see "recognizable, if rather blurred" images. Within a few months he was transmitting the recognizable image of a boy's face. During June of the same year the American Charles Francis Jenkins, who had claimed to transmit images as early as 1923 (which would make them the earliest wireless picture transmissions on record), definitely transmitted the picture of a revolving windmill over five miles between Maryland and Washington. By 1928 Baird not only had transmitted a television signal between London and New York but had demonstrated an experimental color television. By 1929 he had formed the Baird Company, whose transmissions were taken over by the BBC in 1932.

Perhaps the most famous—certainly the most colorful—inventor to participate in the development of mechanical television was none other than Lev Theremin, the legendary creator of the early electronic musical instrument bearing his name (which can be heard in science-fiction films such as *The Day the Earth Stood Still* and the immortal Beach Boys hit "Good Vibrations"). Before Theremin came to the United States to demonstrate his instrument, he was a well-known radio engineer in Russia, and in 1925 he transmitted his first television image shortly after Baird did. His television can still be seen in the Moscow Polytechnical Museum next to Rozing's. (Contrary to the recent documentary film, though, Theremin was not abducted at gunpoint by the KGB from his Manhattan townhouse and thrown into the Soviet gulag. Rather, he seems to have been engaged in low-level industrial espionage for the Soviet government and returned to the USSR of his own free will. Then he was thrown into the gulag.)

And so, if you were surveying the television scene in the late 1920s, you might, as one author did, refer to Zworykin as "that indefatigable worker," or you might not notice him at all. Yet this was a case of the tortoise and the hare. Zworykin was in it for the long haul, and by 1927 he was also working for David Sarnoff, who did not often allow his employees to reveal trade secrets in public.* In 1929, though, Zworykin did unveil a picture tube he called the kinescope. It was a marked advance over previous versions. One of its main features was that it abandoned magnets to deflect the electron beam in favor of charged plates, an idea Zworykin had picked up—and licensed—during a visit to the Parisian lab of Edouard Belin, Fernand Holweck, and their consultant Louis Chevallier.

The kinescope changed the course of television history, but ironically, in a case of industrial treachery, one of Zworykin's closest associates, engineer Gregory Ogloblinsky, was keeping Chevallier informed of everything that was going on at the Westinghouse labs. Without even mentioning Holweck and Belin, Chevallier preemptively filed a patent for the kinescope, claiming improvements that had in fact been made by Zworykin. RCA was forced to buy back Chevallier's patent for one of Zworykin's greatest achievements. This was only one of probably hundreds of patent disputes that emerged during the development of television, the most famous being between RCA and Philo Farnsworth.

Philo Taylor Farnsworth might have been a hero from a Frank Capra movie. A Mormon by birth, he first saw the light of day in a log cabin near Beaver City, Utah, in 1906. When he was twelve his family moved to Idaho, and they lived far enough away from the nearest school that he was forced to ride a horse every day to classes. Philo eventually moved back to Utah when he entered Brigham Young University, but he had to quit after his second year when his father died. According to the testimony of one of his high school teachers at a later patent interference suit brought by RCA, Farnsworth had the idea for an all-electronic television when he was fourteen.

It is true that Farnsworth headed straight toward electronic television without a glance toward mechanical detours; after he left the university he found some financial backers and started his own company to develop his

---

* Westinghouse, General Electric, and RCA were then involved in an extremely complicated joint venture to develop television. David Sarnoff, vice president at RCA, was in charge of the project. RCA, by the way, was formed in 1919 to take over the American Marconi Company.

ideas. He made rapid progress. He worked immediately to develop both an electronic camera, which he called an "image dissector," and an electronic picture tube, both being based on the CRT. On September 7, 1927, before Zworykin unveiled even his kinescope, Farnsworth had transmitted the first all-electronic television image. When Samuel Morse asked "What hath God wrought?" over a telegraph line between Washington and Baltimore, he was demonstrating a working system, ready for commercial production. That was not the case here. Farnsworth's system could scan in only one direction and managed to produce a moving blob of light on his receiving tube. Much work needed to be done, but Farnsworth persisted.

So did Zworykin. He and his engineers were also working on an electronic camera. *Time* magazine's claim that Zworykin was unable to build an all-electronic television system is simply false. By late 1929 the group, now run by the newly formed RCA Victor, was actually broadcasting test transmissions to six homes near Pittsburgh. Zworykin worked a little more slowly than Farnsworth, but ultimately that was not the test.

The two men actually met in 1930. RCA was then in the process of moving its television group to Camden, New Jersey—not Princeton—where the future development of television would take place. In the middle of the move, the company brass sent Zworykin to San Francisco to visit Farnsworth's laboratories. The two men got along quite well; Farnsworth was flattered to have such an illustrious visitor, and his guest was impressed with what he saw. The question for RCA was whether to buy Farnsworth's system. Interestingly, the RCA engineers decided that Farnsworth would be of greater help as a competitor, assuming he continued to receive financial backing. They were convinced that although Farnsworth had done excellent work, Zworykin's system was proving to be better, but if Farnsworth did come up with an improved system, then RCA could afford to buy it later; if they bought it now, they would be obligated to spend a lot of money on developing a line that might not pan out in the end.

Certainly Farnsworth's achievements spurred Zworykin on to greater efforts. By October 1931 Zworykin's group had come up with a new type of TV camera that Zworykin dubbed the "iconoscope." The basic idea behind the iconoscope (and to some extent the image dissector) isn't too difficult to understand. Light from the object being photographed passes through a lens into the iconoscope tube and strikes an array of miniature photoelectric cells—an artificial retina, as Zworykin liked to say. Through the photoelectric effect (chapter 6), the light causes electrons to be given off, leaving that particular element positively charged. The more light, the more positively charged the area becomes. Now an electron beam from a

CRT scans the array of photo elements. When the electrons hit the positively charged area, they discharge it. This electrical discharge, a tiny current, constitutes the signal; the more light that has struck that area, the greater the discharge. Moreover, after the discharge, the element will charge up again if light continues to strike it. In this way the iconoscope can actually store energy, making the device extrasensitive.

At the time no one knew exactly why the iconoscope worked as well as it did, but by 1933 it so eclipsed its predecessors that it signaled the death knell of mechanical television. Although generally considered Zworykin's crowning achievement, it did not go uncontested. Several inventors from Hungary, England, and Canada claimed they had thought of it first, and RCA bought up their patents just to avoid trouble. Kenjiro Takayanagi also claimed that he had applied for a patent based on the same principle before Zworykin. Takayanagi was no amateur. Regarded in Japan as the father of television, he had begun experiments with cathode ray television in 1926 and for a time produced a number of systems as advanced as those in the West. RCA also bought up his patent.

The notable holdout was Farnsworth. During the Great Depression, Sarnoff was exceptional in pouring so much money into television research. Farnsworth Television, on the other hand, was feeling all the hardships of those years, and in 1931 the company was put up for sale. Philco offered to buy it and leave Farnsworth in control of his inventions. Not knowing of the deal, Sarnoff visited Farnsworth's laboratories and offered $100,000 for the enterprise. Farnsworth wasn't there, but his representative declined, saying that RCA was welcome to license Farnsworth's ideas on a royalty basis. "RCA doesn't pay royalties," Sarnoff is said to have infamously replied, "it collects them." After that, the matter moved into the courts.

Matters came to a head when in 1939 RCA engineers Harley Iams and Albert Rose unveiled an improved version of the iconoscope called the "orthicon," which would turn out to be the basis for the television industry of the 1940s and 1950s. RCA's legal department ran across one of Farnsworth's patents from 1933 and realized that it covered several essential aspects of the orthicon. Patent interference suits had been going on between the two companies for a decade (some of which Farnsworth actually lost), but now RCA had to come to terms. The company initiated patent interference proceedings with the U.S. Patent Office, claiming that the iconoscope and orthicon were improvements of a camera tube for which Zworykin had filed a patent application in 1923. The Patent Office didn't buy it. Whether the orthicon was, as some partisans claim, "a Farnsworth invention wearing an RCA name" is highly debatable, but after two years of legal wrangling RCA was forced to license the Farnsworth patent. For the first time in its history, RCA paid royalties.

It was only a temporary reprieve for Farnsworth. Although his company prospered on defense contracts during World War II, it was sold to International Telephone and Telegraph in 1949. Farnsworth himself fell into depression, became alcoholic, suffered nervous breakdowns, and had to submit to electric shock therapy. He died prematurely in 1971.

The television is the archetypal twentieth-century invention. Far more decisively than the lightbulb, it marked the end of the epoch of the lone inventor attempting to change the world and the beginning of the corporate era when large companies poured millions into R&D groups to create new products. When compared to the story of the radio, the development of television seems like a fairly genteel affair, at least for the first fifty years. Almost certainly this is because the evolution was so incremental that no one at the time dared claim to have created the whole shebang. To turn this epic into a David and Goliath story between two rivals, as *Time* magazine has done, cheapens it immensely. The Patent Office did not decide "who invented television"; it ruled on certain specific and highly technical issues. To claim, as certain Internet denizens do, that without Farnsworth the other television pioneers never would have been heard of borders on insanity. Had Einstein not discovered special relativity, Poincaré probably would have nailed it (some say he did); more certainly, had Farnsworth not invented the TV camera, Zworykin would have (he did).

One thing that is forgotten in posthumous turf wars is that ideas, in some sense, are cheap. Just as important in the technological arena is the realization of the idea. The most successful inventions of the past twenty years have been the fax machine, the CD player, the video player, and the cell phone. No one remembers or cares who invented them. In fact, all but the CD player were invented in the United States, but who manufactures the best? It can hardly be denied that Zworykin's tube, because of its storage ability, was a better product than Farnsworth's. True, Farnsworth did it first, but if one wants to call him the inventor of the television, one should not call Edison the inventor of the lightbulb. Others did it earlier. Edison did it better. All of which shows how infantile much of the public debate over these matters is. That *Time* is now regarded as the arbiter of intellectual issues is enough to give one indigestion. Farnsworth and Zworykin would almost certainly feel the same way.

16 / Plausibility: The Invention of Secret Electronic Communication

If a more improbable opening conversation has ever taken place, history has failed to record it. Hollywood screen siren Hedy Lamarr had read bad-boy composer George Antheil's articles on endocrinology in *Esquire* magazine and invited him to meet her at the home of Adrian, the dress designer. As Antheil himself, an admirer of P. T. Barnum, described the encounter in his memoirs:

> They were already sitting at the dining table, one of green onyx splashed with golden tablewear.
>
> I sat down and turned my eyes upon Hedy Lamarr. My eyeballs sizzled, but I could not take them away. Here, undoubtedly, was the most beautiful woman on earth. Most movie queens don't look so good when you see them in the flesh, but this one looked better, infinitely better than on the screen. Her breasts were fine too, real postpituitary.
>
> The black silken ringlets fell softly down around her throat, and . . . oh well, why go on? You can get the same effect by going to your favorite movie theater and pretend you're looking across the dinner table, just like lucky me.
>
> And—remember!—this picture is in technicolor.
>
> So I looked at her and looked at her, and finally I permitted my eyes to look down a little from her face. I felt a terrible flush spreading over my map.
>
> "But your breasts," I stuttered, "your breasts—"
>
> I could not go on.
>
> She whipped out a notebook and a pencil. "Yes, yes," she said breathlessly, "my breasts?"
>
> "Your breasts . . ." I repeated aimlessly, but my mind commenced to wander. I could not go on. I knew that in a moment I would swoon, but Adrian shoved a glass of water into my hand just in the nick of time. I wolfed it and said:
>
> "They are too small." (I just said that to lead her on; every movie star wants larger bosoms.)
>
> Hedy made a note in her book. "Go on," she said, not unkindly.
>
> "Well," I said, wanting to get up and rush right out of the United States, "well, they don't really *have* to be, you know."

She made another note, taking some time to do it. The butler took away my untouched hors d'oeuvres. Silence reigned, and I knew that more was expected of me.

"You are a thymocentric, of the anterior-pituitary variety, what I call a 'prepit-thymus,'" I volunteered.

Hedy Lamarr kept on writing for a moment, and then said, "I know it, I've studied your charts in *Esquire*. Now what I want to know is, what shall I do about it? Adrian says you're wonderful . . ."

"Well," I said, "your breasts . . . they . . . so to speak . . . if you're short on postpituitary . . . the thing to do is . . . er, activating substance . . . breasts can be controlled by. . . ."

(Oh, God, I wanted to *die* of shame.)

"Go on, go on," Hedy said, becoming a bit restless. "The thing is, can they be made *bigger?*"

"Yes," I said, "much, much bigger!"

"Bigger than this—" I was afraid for a moment to look, but saw that she did not intend to take off her beautiful Hungarian blouse. She was just thrusting out her chest.

"Yes, yes, yes, yes, YES!" I cried.

And that, no kidding, is the way I first got acquainted with our very good friend, Hedy Lamarr.

And that, according to legend, was the origin of all secret electronic communication.

In the year 1900, several months before Max Planck invented quantum mechanics, George Antheil was born in Trenton, New Jersey, across from the state penitentiary. The proximity to the prison proved fortuitous: According to Antheil, his interest in music was first stimulated by "two old maids" who played the piano next door, day and night—a cover, it turned out, for "one of the most sensational prison breaks" in Trenton's history. Thus launched, Antheil developed into a child prodigy who seemed destined for a career as a concert pianist, and at age twenty-two he set sail for Europe, fame, and fortune. Possibly. Antheil assures us he went to Europe not in pursuit of a concert career but in pursuit of a fiancée, a "well-edited version of Lana Turner and Betty Grable combined," who had been spirited off by her parents when they discovered her ill-advised plans to marry the musician. Though desperately in love, Antheil soon forgot the girl and became notorious for his concerts, which invariably ended with a few of his own "ultramodern" works, followed by a riot. Antheil—it is an indisputable fact—carried a .32 automatic in a silk shoulder holster in order to fight his way out of the concert hall should it become necessary. Whether he ever shot anyone remains unclear.

The public did not riot because Antheil's pieces were bad, though they may have thought so. A composer since childhood, Antheil had studied at the Settlement School of Music in Philadelphia (though he claimed it was at the Curtis Institute) and with Ernest Bloch in New York, and by the time he reached Europe, he had assimilated all the current trends of modernism. Then he went beyond them. His "mechanisms," with titles such as "Mechanisms," "Airplane Sonata," and "Death of the Machines," were at the forefront of the avant-garde, and within a few seasons he was anointed the "first composer of his generation," which he remained for about three years. In those days artists believed that progress was possible, and audiences had yet to succumb to the fathomless boredom made possible by twentieth-century technology, with its capacity for infinite repetition. Riots were the order of the day, and nothing could be wished for more than a succès de scandale.

Inevitably Antheil settled in Paris. In 1923, deciding to devote himself fully to composition, he moved with his Hungarian fiancée, Boski, into a tiny garret above Sylvia Beach's famous bookstore, Shakespeare and Company. The poverty was desperate, but the friends were James Joyce, Jean Cocteau, Ernest Hemingway, T. S. Eliot, and, above all, Ezra Pound. Their alliance proved mutually beneficial. Apart from his activities as poet and critic, Pound considered himself a composer and wrote a one-act opera, *Le Testament de Villon,* in a neoantique, Stravinskian style. Whether Antheil merely "re-rited" it, in Pound's words, or made it "virtually unperformable," in the words of Pound adherents, is the subject of hot debate, but the opera has been performed. Work on it stimulated Pound to buy a bassoon and annoy Hemingway.

To return the favor, Pound, then kingmaker in Parisian artistic circles, wrote a book, *Antheil and the Treatise on Harmony,* which was meant to propagandize his ideas on music and his new friend. Whether Antheil coauthored it or was merely embarrassed by it also remains the subject of internecine strife. Surprisingly for a work on music, it is filled with references to machines and the fourth dimension:

> Antheil is probably the first artist to use machines, I mean actual modern machines, without bathos.
>
> Just as Picasso, and Lewis, and Brancusi have made us increasingly aware of form . . . so Antheil is making his hearers increasingly aware of time-space, the division of time-space.
>
> Antheil is supremely sensitive to the existence of music in time-space. The use of the term "fourth dimension" is probably as confusing

in Einstein as in Antheil. I believe that Einstein is capable of conceiving the factor time as affecting space relations. He does this in a mode hitherto little used, and with certain quirks that had not been used by engineers before him; though the time element enters into engineering computations.

Machines? Time-space? The fourth dimension? What does this have to do with Antheil, secret communications, and, above all, Hedy Lamarr? As it turns out, everything.

Since the mid-nineteenthth century, when the mathematician Georg Bernhard Riemann (1826–1866) had introduced the idea of higher-dimensional, non-Euclidean geometries, popularizers, including Hermann von Helmholtz, Henri Poincaré, and Lewis Carroll, had exposed the public to the idea of the possibility of the fourth dimension. Speculation on the nature of this fourth dimension became the rage in Europe, producing, among other things, at least one "hyperspace philosopher" and Edwin Abbott's famous novel *Flatland: A Romance of Many Dimensions.* Indeed, fascination with the fourth dimension eclipsed the black hole mania that peaked in America around 1980, and reached such a pitch that in 1909 *Scientific American* sponsored an essay contest for the best popular explanation of the fourth dimension.

In those days, while Einstein was still an unknown patent clerk, the fourth dimension was conceived to be another spatial dimension, which humans were unable to perceive, much like the six extra spatial dimensions of today's superstring theories. (The single exception to this trend was H. G. Wells and his prophetic 1895 novel *The Time Machine.*) Artists, as always, latched on to the latest scientific speculations, and the notion that we really live in a four-dimensional world profoundly influenced the Cubists, the Italian Futurists, the Russian Supremacists, and later the German Bauhaus architects, all of whom attempted to portray their subjects from a multidimensional perspective.

In 1919 everything changed. After the eclipse expedition that made Einstein world famous (chapter 8), the fourth dimension became, for now and until eternity, time. Antheil and Pound, at the avant of the avant-garde, were not to be left behind in applying this new concept to music. As early as 1922 Antheil wrote in the Dutch art journal *De Stijl:*

> My forms are the first complete forms that have come out of the only forms out of which musical forms can be made . . . TIME. Is not TIME, and TIME ALONE the SOLE canvas of music? . . .
>
> Now I hope to present you not with an explosion, but with the FOURTH DIMENSION . . . THE FIRST PHYSICAL REALIZATION OF THE FOURTH DIMENSION.

Here we encounter the Panopticon's best artifact in the permanent exhibit on Misunderstanding in the Aid of Progress. Antheil's understanding of relativity may have been on a par with Deepak Chopra's understanding of quantum mechanics, but he remained undeterred in his attempts to incorporate relativity theory into music. How was one to do such a thing? For Pound, Antheil, and colleagues, the answer was obvious: machines. Machines operate in time as well as space and therefore represented the ideal medium to express the unfolding of time-space. "Machines acting in time space," wrote Pound, "and hardly existing save when in action, belong chiefly to an art acting in time space. Antheil has used them effectively. That is a *fait accompli* and the academicians can worry over it if they like."

Later Antheil denied that his works attempted to portray machines, but early on it was otherwise: "I saw thousands of electric lamps strung in the heavens and illuminated from one switchboard to create God; vast cinemas projected a new dimension in the skies; music machines large enough to vibrate whole cities. All these although later appropriated were first my very own. The ecstatic poetry of space! The satisfying hardness of time!"

The ecstasy of machines. Relativity, machines, time-space. All these influences converged in the artists of the the age to result in the machine aesthetic so characteristic of the 1920s. In music, Antheil was demonstrably first, despite the fact that Honegger, Mosolov, and Prokofiev received the greater accolades. In painting, it was Fernand Léger who epitomized the mechanization of the age, and so, inevitably, the two revolutionaries would combine forces, with Man Ray and Dudley Murphy, to produce the first abstract film, the *Ballet Mécanique*.

Antheil's original 1925 score for the short film called for sixteen synchronized player pianos. With the technology of the day, such a feat proved impossible, and he rewrote the score for a single player piano and an assortment of percussion, sirens, buzzers, and an airplane propeller. Still, the music turned out to be twice as long as the film, and so Léger's and Antheil's creations assumed separate lives. In its Paris and New York premieres, Antheil's *Ballet Mécanique,* with its complete overthrow of traditional instrumentation, naturally provoked riots. But it remains an indisputable masterpiece, Antheil's best work, and the first concert work to incorporate a player piano.

The *Ballet* proved to be the pinnacle of Antheil's career, in terms of vision, quality, and notoriety. From then on his music became progressively more conservative and less distinguished. By the early 1930s he was out of fashion, broke as usual. Neither did the situation in Europe appear promising: Pound no longer championed him, and Hitler was rising to power. In 1933 Antheil decided to return to the United States.

But Antheil, with his facile if not very deep mind, was a hard man to corner. In his perpetual attempts to get rich, he patented a revolutionary "See Note" system, an instantaneous method for learning piano. Eschewing conventional notation, Antheil ruled paper to resemble an actual piano keyboard; the position of the notes was where you would place your fingers, and a note's duration was indicated by its actual length on the paper. One read not across but downward. In a word, Antheil's system resembled nothing so much as a player-piano roll, and the notation, like player-piano holes, was essentially digital: on-off, on-off. He published a pair of articles about his revolutionary system in 1938, in the newly founded *Esquire* magazine, but the revolution flopped utterly and completely. Aside from the sheer unlikelihood of such an approach overthrowing tradition, Antheil had forgotten an essential ingredient: compatibility. The system might have worked for piano. What if you played the bassoon?

See Note was a minor diversion. Years earlier in Berlin, he tells us, a roommate had bequeathed him some books on endocrinology by the Columbia University professor Louis Berman and, starving and with nothing else substantial to nourish him, Antheil fed ravenously. Starvation is a wonderful thing; how lucid it makes the mind. Berman's theory was simple: Hormones are destiny. So taken was Antheil by this simple reduction that he wrote a book himself, *Every Man His Own Detective: A Study in Glandular Criminology,* which was published in 1937. It explained how an "endocrine criminologist" visiting a crime scene could by surveying the forensic evidence immediately determine what hormonal type had perpetrated the crime. The thymocentric is the most dangerous type—over 70 percent of all prisoners in our penitentiaries (especially Trenton's) are regulated by the thymus gland, and therefore we should spend 70 percent of our time studying the thymocentric, who is usually hyphenated. To elucidate, the other endocrine glands come into play, and so we have the thymocentric-pituitary, the thymocentric-adrenal, and so on. Remember, the thymo is a bad *hombre* and always shoots and knifes in the back. Robert James, however, that human devil who bound his wife to a table, thrust her legs into a cage filled with rattlesnakes, and threw her still alive into a garden pool, was a pituitocentric.

So serious was Antheil's belief in endocrinology that, according to one unimpeachable rumor, the Parisian police made him an honorary lifetime member. Around 1930, under the pseudonym Stacey Bishop, he wrote a detective novel, *Death in the Dark.* Antheil assures us that the manuscript was edited on the Italian Riviera by Pound, Eliot, Yeats, and Gerhart Hauptmann; it was inexplicably published by Eliot at Faber and Faber. Although the mystery in *Death in the Dark* may be solved by endocrinological means, the book remains virtually unreadable, which shows the wisdom of writing by committee, even when two of the committee members have won the Nobel prize.

The most visible of Antheil's writings on endocrinology, however, were the articles he published in *Esquire* in 1936. These pieces, entitled "Glands on a Hobby Horse," "Glandbook for the Questing Male," and "The Glandbook in Practical Use," did not concern criminology. They concerned how to pick up women. By today's standards they are strangely premonitory—witness current ads for hormonal breast-enhancement cream. They are also perhaps not in as bad taste as Jerry Springer, advising the reader that the "type A" woman is dominated by the postpituitary (excessive) and rating her accessibility as extremely high with a strong tendency to nymphomania. The charts, ranging from A to D with subclasses, detail all "outstanding characteristics at twenty feet": walk, build, height, and bust.

At about the same time the articles appeared, Antheil, in his perpetual state of impoverishment, made the ultimate sacrifice and moved to Hollywood to write movie music. He remained in la-la land for the rest of his life, productively if anonymously, writing the scores for any number of films. Cecile B. De Mille's *Plainsman* is probably the most famous. Catch the ferocious war dance as the Indians roast Gary Cooper on late-night TV.

Hollywood did have its compensations: By 1940 and the outbreak of World War II, Antheil's *Esquire* articles had fallen into the hands of a young Hollywood superstar who was desperate to increase her bustline.

Hedy Lamarr's career was shorter but scarcely less eccentric than Antheil's. An Austrian, Hedwig Eva Maria Kiesler was born in Vienna, in 1915, 1914, or 1913, depending on whom you believe. By the age of fourteen (fifteen, sixteen), she had decided on a theatrical career, her parents granted the request, and she played a few bit parts in films shot at the local studio. Afterward she briefly attended Max Reinhardt's famous dramatic school in Berlin. The turning point in her life came in 1932, when she was chosen to star in the Czechoslovakian silent film *Ecstasy,* which included cinema's first, or at least most notorious, nude scene—a few seconds of Hedy scampering through a glade and diving into a sylvan pool. According to her autobiography, *Ecstasy and Me,* she was tricked into doing the scene; the director had agreed to place the cameras on a distant hill but had not told her that he would be using a telephoto lens. It should also be pointed out that she later sued the publisher of *Ecstasy and Me* for $23 million, claiming she did not write it. Suing anyone who mentioned her in public became a lifelong habit.

The movie made her notorious. By the time *Ecstasy* was released in 1933, her great beauty—certainly not her acting—had already brought her to the eye of Fritz Mandl, who courted her and married her the same year. Mandl was the owner of Hirtenberger Patronenfabrik, Austria's leading munitions manufacturer, and a Nazi sympathizer. Hans-Joachim Braun, writing for *American Heritage of Invention and Technology*, describes the nature of Mandl's dealings. Like Antheil, he was born in 1900. When the Treaty of Versailles after World War I forbade weapons manufacture in Austria, Mandl (who had by then taken over his father's business) set up subsidiaries in Poland, the Netherlands, and Switzerland. When he was caught selling weapons to Hungary in contravention of the treaty, he turned around and began selling them to Spain during that country's civil war. So willing was he to do business with anyone, anywhere, and at any time that the Nazis confiscated his firm even before the Anschluss of 1938, which joined Austria to Germany. At that point he moved to Argentina and became an advisor to Juan Perón.

One must question Hedy's judgment in marrying such a man, which required her to host Hitler and Mussolini, but she was only seventeen (eighteen, ninteen) at the time. Despite the palatial splendor in which she lived, she appears to have been somewhat terrified of her husband. At one point she tried to sneak out on her own, but Mandl followed. Hedy ducked into an empty room in a brothel to escape him; a customer entered as Mandl was heard pounding on the door behind her. By her own admission, Hedy had always been oversexed, and rather than face Mandl's wrath, she obliged the customer. (On the other hand, she did sue the publisher.)

The marriage had its bright spots. Mandl spent two years trying to buy up every print of *Ecstasy* in existence, but he also allowed his young wife to sit in on the firm's business and planning sessions, which included design of aircraft control-guidance systems. He may have thought she understood nothing, but apart from being "the most beautiful woman who ever graced the silver screen," she was a smart, if kooky, cookie. In spite of such romantic interludes, the marriage did not last long. In 1937 Hedy hired a new maid who resembled her, one day drugged her, and, disguised in the maid's uniform, fled to Paris. Then she divorced Mandl on grounds of desertion.

From Paris, Hedy Kiesler went on to London, where she met Louis B. Mayer of MGM. After remarking that "her chest was bigger than he thought," he offered her a job at his Hollywood studio for $125 a week. He renamed her Hedy Lamarr.

The night following their initial encounter, George met Hedy at her mansion, and this time the discussion turned to the war. Hedy, already one of Hollywood's biggest stars, was nevertheless unhappy making a fortune while the world she knew was in flames. She let drop that she was thinking of quitting MGM and moving to Washington, D.C., where she would offer her services to the newly established National Inventor's Council. "They could just have me around," she volunteered, "and ask me questions."

Naturally, Antheil wanted to know what good that would do, and now Hedy revealed that she was good at inventing weapons. One of her ideas was for a torpedo guidance system. When you launch a torpedo at an enemy ship, you want to be able to guide it to the target; at the same time you do not want the enemy to be able to intercept the missile and divert it—or turn it around. You need a secure guidance system. Hedy called her proposal "frequency hopping": If you had a transmitter that shifted frequencies in a prescribed, pseudorandom fashion, and a receiver hopping around in synchrony with it, then the receiver could pick up the guidance instructions without difficulty, but any surreptitious eavesdropping would be impossible. An enemy listening in on any one frequency would only hear a blip as the signal passed through that band.

Hedy did not have the necessary technical background to put the idea into practice. Neither did George, exactly, but he immediately saw the solution: player pianos. The transmitter and receiver could be synchronized by controlling them with two player-piano rolls, punched with identical random patterns of holes. He even proposed that the system utilize eighty-eight different frequencies—the number of keys on a piano. Finally he had found a use for his See Note system!

For several months they discussed their invention. In one version Antheil himself decided they should patent it. In another Charles Kettering, the director of General Motors' research division, who then headed the Inventor's Council, made the suggestion. Then, with the help of a Professor MacKowen of electrical engineering at Caltech, Antheil ironed out the bugs, and on June 10, 1941, Hedy Markey (by then she was on husband number two) and George Antheil filed a patent application. A year later, on August 11, 1942, the two inventors received U.S. patent number 2,292,387 for a "secret communication system."

Never one to miss out an opportunity for self-promotion, Antheil spent considerable effort lobbying the navy to adopt their invention, but the navy turned it down on the grounds that it would be too bulky. Evidently they thought Antheil wanted to put a player piano in a torpedo.

Thus the Lamarr-Antheil guidance system was shelved and failed to alter the course of the war. Antheil died in obscurity of a heart attack in 1959 and never saw the fruits of his labors. Lamarr's screen career faded rapidly; she ran through six husbands and several fortunes altogether and was arrested twice on shoplifting charges. (Despite strenuous denials in her autobiography, acquaintances confirm she was a kleptomaniac.) Without work, she continued to sue virtually anyone who mentioned her in public. In 1998 she took action against the Canadian software firm Corel, which, thinking she was dead, had used her likeness on the box for their CorelDraw8. She sought $15 million in damages; they settled for less and made her a consultant. Hedy Lamarr died in Florida, finally well off but something of a recluse, on January 19, 2000. A suit against the Gallo winery was in litigation.

Despite the bittersweet fate of the protagonists, the tale of frequency hopping has a happy ending. Ten years after navy brass dismissed Antheil as a nutcase, electronics had improved to the point of being able to implement frequency hopping, and the Lamarr-Antheil patent became the basis for all secret military communications. By the time of the Cuban missile crisis the armed forces were routinely using frequency hopping to scramble signals.

That is not all. Frequency hopping, today known as spread-spectrum technology because it spreads the signal over a large part of the electromagnetic spectrum rather than confining it to a narrow band, has other advantages. It is difficult to jam spread-spectrum broadcasts because even the most intense jamming signal on one frequency will interfere with only a tiny portion of the full transmission. Furthermore, if the transmitted signal is hopping all over the place, most of the frequencies at any instant of time are left unused. Thus spread spectrum is ideal for interleaving (multiplexing) many messages simultaneously, and it becomes a more efficient transmitting method than ordinary single-frequency techniques. For those reasons, spread spectrum is now becoming the basis of Internet and cell phone traffic. And at the bottom of it all lie some crazy ideas about relativity theory, player pianos, endocrine glands, and a movie star who enjoyed designing weapons.

Such a fabulously improbable yarn could be little else than true. And it is—as far as it goes. The patent exists and is readily available on the Internet. Thousands of tributes to Hedy Lamarr may also be found there. She and Antheil (posthumously) received a special Electronic Frontier Foundation Pioneer Award in 1997. When told of the award, she replied, "It's about time." Tabloids proclaimed her the genius behind the Internet.

The urban legend is by now immutable, but as usual with such a tale—even such a great one—there is more than meets the eye. It's not entirely clear how much Hedy understood about her invention. In a somewhat tongue-in-cheek interview published in the armed forces newspaper *Stars and Stripes* Hedy declares that Antheil "did the really important chemical part." It was fun to watch "them" (Antheil and the engineer?) "put together all the little thingamabobs that went into the device." Lamarr never mentions the invention or Antheil in her autobiography, which admittedly is more concerned with her sexual adventures. She did, however, disclose in the *Stars and Stripes* interview that "it was lots more fun being scientific than going to the movies."

Hans-Joachim Braun, the author of the *American Heritage* article, has also informed me that he has found a document in the Bundesarchiv/Militararchiv in Freiburg proving the subject of frequency shifting was discussed in July 1939 at the German firm of Siemens and Halske; Braun assumes that it probably came up at Mandl's firm several years earlier. One can't dismiss the possibility that Hedy was merely passing on the contents of a planning session that she had overheard. Robert Price, a retired MIT engineer and an authority on the history of secret communications, at one time interviewed Lamarr. Mandl, he reports, was a Jew, a Jew selling arms to Hitler. Most of Lamarr's circle were Nazi sympathizers, but she had become vehmently anti-Nazi. Price feels certain that she was basically acting as a spy—"the Mata Hari of World War II"—stealing the secrets of her husband's firm and conveying them to the West.

On the other hand, Antheil always gave her the greater part of the credit, and in a letter to a Mr. Reynolds, apparently the director of the National Inventor's Council, he says:

> Likewise, a curiosity of this idea is that its co-inventor is Miss Hedy Lamarr, the motion picture actress (who is a good friend of mine), who, curiously enough, has had considerable experience of a second-hand nature concerning this subject. Her first husband, Fritz Mendel [*sic*], was once one of the largest munition manufacturers in Austria, besides which Miss Lamarr has a natural aptitude for the rather unfeminine occupation of inventor.

Is Antheil implying that the idea originated with her or her husband? Hmm. The letter actually concerned *another* invention on which the two were collaborating: a "magnetic antiaircraft shell," which was supposed to detect the proximity of an enemy airplane by a "magnetic device" and cause the shell to explode on target. The workings of this "magnetic device," however, are simply not explained, leading one to wonder whether the inventors knew what they were talking about. Antheil does offer to send a more elaborate description and requests a prompt response for an unusual reason:

> Inasmuch as I am very happily—and very suitably married—and also inasmuch as no young married wife can long endure the story that Miss Hedy Lamarr and myself are working upon an anti-aircraft shell together (which is the story which Miss Lamarr—truthfully—tells my wife!) some degree of haste is required if this family, which include a three and one half year old son, is to be kept together!

It does not appear that the magnetic antiaircraft shell got very far; judging from a 1941 letter of Antheil to Lamarr, he gave up the project when Hedy became suspicious that he was trying to defraud her. Perhaps she threatened to sue.

So who did what? We will never know. We do know, however, that the concept of frequency hopping has had a long history. David Kahn, author of *The Codebreakers,* writes in his article "Cryptology and the Origins of Spread Spectrum" that in 1929 a Polish engineer, Leonard Danilewicz, proposed to the Polish army a system for secret radio telegraphy, which, he later mourned, "unfortunately did not win acceptance, as it was a truly barbaric idea consisting of constant changes of transmitter frequency." In the 1930s a Swiss inventor, Gustav Guanella, proposed a similar idea, and in 1935 two Telefunken engineers, Paul Kotowski and Kurt Dannehl, applied for a patent for a device to hide voice signals under a "broadband noiselike signal produced by a rotating generator."

During World War II spread-spectrum devices were already in action, on both sides. They were used mostly in radar, where synchronization of the transmitter and receiver is not a problem (because transmitter and receiver are at the same location). The most famous use of frequency hopping during the war was the ultrasecret SIGSALY system, which in 1944 scrambled the telephone conversations between Franklin Roosevelt and Winston Churchill.* It was the first absolutely unbreakable scrambling system. SIGSALY's workings were far too complex to describe in detail here. Roughly speaking, SIGSALY first sampled the amplitude level (loudness) of Churchill and Roosevelt's voices and "quantized" them. Today we would say the system effectively digitized the voices. It next added a randomly generated number to each sample, scrambling the voice levels. The

---

* The Army Signal Corps prefaced all its code names with SIG, but the remainder was apparently randomly generated; SIGSALY is not an acronym.

now random intensities were broadcast across the Atlantic by FM radio, which converts every amplitude level to a different frequency. Because all this took place in a totally unpredictable fashion, the message was impossible to crack.

How were the random numbers generated? Mercury vapor lamps produce perfectly random "white noise," or static. Most engineers try to supress noise; the SIGSALY designers used it. Fifty times a second they sampled the noise levels and recorded them onto vinyl disks, essentially record masters. Vinyl, then, encoded the random digits later added to the voice signals. The disks played the same role as Antheil's player piano rolls; one was sent to Washington and another to London under high security. Synchronization between the Washington and London disks was accomplished by a quartz-crystal oscillator, much like those found in Swiss watches today, except that these were housed in seven-foot-tall ovens shaped like horseshoes. Independently tuned to the National Bureau of Standards' timing signal, they started up the two SIGSALY stations at the designated instant for the conversation. The disks were destroyed after a single usage. SIGSALY, one must acknowledge, was a bit more sophisticated than player-piano rolls. The SIGSALY designers in fact implemented no fewer than eight major innovations in electronic communication. Price considers it the greatest technological achievement of World War II, after the atomic bomb and radar.

Following the war, spread spectrum continued to be developed at MIT, and by 1947 engineers at Sylvania were using spread spectrum to guide missiles. At about the same time Mortimer Rogoff, at the Federal Telecommunication Laboratories in Nutley, New Jersey, used the Manhattan telephone directory to generate a pseudorandom key. By 1955 spread spectrum was overcoming jamming.

This, more plausibly, is the true evolutionary trunk of spread-spectrum technology. The fact is, secret communication was invented in secret, and that a movie star has become enshrined as its originator is a bit of only-in-America hockum. Price, who worked at MIT on secret communications systems in the 1950s—with no knowledge of Lamarr and Antheil—also points out that civilian communications systems use a variation of spread spectrum known as code division multiple access. When International Telephone and Telegraph went to patent CDMA in the mid-1950s, ITT's researchers discovered the Lamarr-Antheil patent as dead prior art but duly cited it (and Hedy demanded royalties). This, in the technical world at least, was the resurrection of the Lamarr-Antheil invention. One can readily agree with David Kahn, who concludes, "Though only a sidelight in the history of spread spectrum, because it had no direct influence on the evolution of the technology, the frequency-hopping invention did impart to that field its most glittering bit of glamour."

That peroration may be a bit of a letdown, but there are several lessons to be learned from the tale of Hedy and George, most of them cheerful. Scientists too often sneer at attempts by artists to find meaning in the great theories of nature. How is it possible to translate special relativity into music? Perhaps it isn't, but in the twentieth-century artists' attempts to interpret the fourth dimension, we have a beautiful example of a situation in which naive efforts profoundly influenced the entire course of modern art. In the case of one brash young composer, his misunderstandings about spacetime led in part to a significant invention. If these naive artists had, as many scientists wish, decided that science lay forever outside their province, civilization would have been, and would be, impoverished. One hopes the Lamarr-Antheil tale inculcates a greater live-and-let-live attitude toward amateurs. They may not always get it right, but sometimes to get it wrong is better.

Scientists should foster a more indulgent attitude toward mavericks, but the reverse also applies. The general public romanticizes the maverick at the expense of the less glamorous trench soldier. In the iconification of the Lamarr-Antheil patent as the basis of all secret communication we see the same adoration of the outsider as was bestowed on Buckminster Fuller. To this day virtually everyone believes Fuller invented the geodesic dome, but as mentioned in the introduction to the Technology Domain, it is a matter of record that a more conventional engineer did it thirty years earlier.

What is forgotten in all this, as usual, is the Panopticon's dusty exhibit on Progress as Perspiration. As we have seen throughout, scientific and technological progress tends to be incremental. Movie stars may have brilliant ideas, but it takes the labor of an unknown composer to bring it to fruition. Andrew Wiles, who proved Fermat's Last Theorem, once remarked to me that he could explain the idea behind his celebrated proof to another specialist in three or four minutes. But it took Wiles eight years to work out the consequences. He also confessed to me that he "tried everything," like the legendary monkeys who type out *War and Peace* by trial and error. At the same time, he was quite aware that "the person who carries the torch past the finish line gets all the credit." A prophetic remark.

Before the finish, there is inspiration, trial and error, incremental progress, and misunderstandings. All are required to carry the race forward, and all should be respected.

III

The Domain of Chemistry and Biology

We now abandon the realm of steel and electricity, laws of motion, and quantum mechanics for the softer realms of liquids, gases, flesh, and sinew. The lessons found in the Panoptican's Domain of Chemistry and Biology are perhaps no different from those discovered on the previous perambulations, except that in days gone by natural philosophers with inclinations toward chemistry and biology seem to have lived more exciting lives than those who devoted themselves exclusively to physics. One wonders whether this had something to do with a social conscience.

An excellent example is Joseph Priestley (1703–1804), generally mentioned in chemistry texts as having discovered oxygen in 1774. In this he was preceded by the Swede Carl Wilhelm Scheele, but unfortunately Scheele's book was not published until 1777, by which time Priestley's discovery was well known. Despite his renown, however, Priestley's life took a dramatic and dangerous course. In the decade following his work on oxygen the liberal Priestley became ever more involved in religious and political controversy, displaying open sympathy with the French revolutionaries, which earned him the enmity of English conservatives, notably Edmund Burke. Events reached a head on the second anniversary of the storming of the Bastille, when riots broke out in Birmingham, Priestley's home. His house and laboratory were destroyed, and family members were forced to flee for their lives to London. Finding the atmosphere there scarcely less hostile, Priestley emigrated to America and spent the last ten years of his life writing in Northumberland, Pennsylvania. Priestley's fate was certainly not lost on William Lawrence, whom we shall encounter shortly.

The French Revolution and oxygen were also perhaps the two most important ingredients in the life of Antoine Lavoisier (1743–1794), who is often accorded the title of "father of modern chemistry." Lavoisier's great fame rests largely on the fact that he overturned the infamous phlogiston theory and, in doing so, explained combustion. For a full century natural philosophers had been trying to understand what was really happening when a substance burned. Most substances undergoing combustion— wood, for example—lose weight. The remaining ash weighs only a fraction of the original log. The question was, where did everything go? As early as 1667 Johann Becher proposed that combustible substances were rich in *terra pinguis,* which was lost during combustion. His disciple George Ernst Stahl replaced *terra pinguis* by the mysterious phlogiston, from the Greek for "inflammable," and posited that it was phlogiston that vanished during burning. The trouble was, among other things, that some substances actually gain weight under combustion (as iron does when it rusts). In this case the phlogiston needed to have negative weight.

Lavoisier, somewhat grudgingly acknowledging the help he got from Priestley, demonstrated in a series of controlled experiments that combustion of a substance is its combination with oxygen (a term Lavoisier coined). He also showed that water was formed by the combustion of hydrogen and that therefore water was composed of oxygen and hydrogen. However, Lavoisier had published his paper on the subject after learning about experiments along these lines performed by Henry Cavendish, accused by James Watt of having plagiarized the results from *him.* It appears, though, that Cavendish had learned about it from Priestley, who had noticed that igniting hydrogen and oxygen with an electric spark (an idea Volta suggested to him) left a film of dew on the walls of the vessels in which the hydrogen was burned. Gaspard Monge, the mathematician, had performed similar experiments at the same time, and the five-way priority dispute, as intense as any other we have witnessed, has not ended to this day. P. J. Macquer, who noticed the dew on the vessels before anyone else, slipped by anonymously.

In 1789 Lavoisier published all his results in his *Traité elementaire de chimie,* which is to chemistry what Newton's *Principia* is to physics. It contained the first "modern" table of the elements; it announced a new theory of heat, termed by Lavoisier "caloric"; and it announced the law of conservation of matter. Did it? Most scientists say so, but the law is rather tacitly assumed rather than expressed outright. He says, for instance, "As no substance can furnish a product larger than its original bulk, it follows, that something else has united with the alcohol during the combustion." In any case, here Lavoisier had been anticipated by three decades by the Russian polymath Mikhail Vasilievich Lomonosov (1711–1765). Lomonosov, the Russian Ben Franklin, not only duplicated his American counterpart's thunderstorm experiments but wrote histories, poetry, and the first

Russian grammar (often taken to be the beginning of modern Russian), made contributions to chemistry and physics, and helped found Moscow University, which is now named after him.* Lomonosov, an early foe of the phlogiston theory, wrote to the great mathematician Leonard Euler in 1748, "All changes occurring in Nature are subject to the condition that, if so much is taken away from one substance, just as much is added to another." In 1750 he submitted such views to the Academy of Sciences, and at least one historian has claimed that Lavoisier read Lomonosov's papers without acknowledgment.

Despite his condescension toward predecessors and contemporaries, Lavoisier's fate was undeservedly tragic: A member of the hated Ferme Générale, the institution responsible for tax collection, Lavoisier became the target of absurd attacks made by Jean-Paul Marat during the Reign of Terror and was guillotined.

Strangely intertwined with Lavoisier and caloric was the career of Benjamin Thompson. If introductory textbooks mention him at all these days, it is usually a laconic line to the effect that Lavoisier's caloric theory was disproved by "Benjamin Thompson, an American who later became Count Rumford of Bavaria." More laconically, he was a bastard from beginning to end. Massachusetts-born in 1753 with a genetically enhanced gift for social climbing, Thompson at the age of nineteen married thirty-three-year-old Sarah Rolfe of New Hampshire, the rich widow of Colonel Benjamin Rolfe. Governor John Wentworth promptly took Thompson under his wing, sent him on surveying expeditions, and taught him science and philosophy. Politics as well. Major Thompson of the New Hampshire militia was soon engaged in an elaborate undercover operation to forcibly return British army deserters to their units. When the scheme came to light, Concord's Committee of Correspondence charged Thompson with being a "rebel to the state"; on the verge of being tarred and feathered, Thompson fled to Massachusetts, abandoning his pregnant wife to the angry mob. He never saw her again, and never inquired after her until thirty years later, when he wanted to remarry. By then she was dead.

* During one thunderstorm Lomonosov and a friend, Professor Richmann, were simultaneously carrying out experiments with what amounted to lightning rods. Lomonosov was nearly killed. Suddenly Richmann's servant burst into the house, informing him that Richmann *had* been killed. Lomonosov was devastated at the loss of his friend but managed to write, "Nonetheless, Mr. Richmann died a splendid death, fulfilling a duty of his profession."

In Boston, Thompson spied on the inner circle of revolutionaries and passed to the British command the Revolutionary War's first known invisible-ink message, which described the Colonials' military plans.* Arrested shortly thereafter, he was freed for lack of evidence and continued to spy for the British until he left the country in October 1775, probably smuggling out papers of the agent Benjamin Church. Thompson did return to America—as Lieutenant Colonel Thompson, fighting for the British. Whether his sudden departure from England was due to suspicions that he was a spy for the French is unclear, but once on these shores he became the most formidable adversary of General Francis Marion, the Swamp Fox. He attempted to capture General Nathanael Greene but failed and so cannot be held responsible for the invention of the cotton gin. Made commander of the King's American Dragoons, Thompson headed north, raised a unit at his own expense, and spent the remainder of the war ravaging the countryside of Long Island.

Inevitably, Thompson was knighted. At the same time King George granted him permission to enter the service of the duke of Bavaria, with the understanding that he would be passing any useful political information to England. (This activity does not seem to have lasted long.) The elector of Bavaria put Thompson in charge of the complete reorganization of the country's military, and for this service he was made Graf von Rumford. It was in Bavaria that Thompson carried out the experiments that made him immortal. This actually did not represent a sudden philosophical turn on behalf of the amoral soldier of fortune; Thompson had always puttered around in his spare time and at one point even designed a ship. After arriving in Bavaria he studied the properties of gases, colors, and the action of light on silk; he improved cookery and ovens, essentially invented the modern fireplace, and investigated the nature of heat.

Lavoisier had looked upon heat as a fluid, caloric, which flowed from hot bodies to cold bodies. It is not a bad model at all, but if true, if you place a hot object atop a cold object with an insulator in between, there should be no temperature change in either (due to their proximity), since the caloric fluid cannot flow from the one to the other. Precisely this experiment with hot water and ice convinced Thompson in 1797 that heat was not fluid but he motion of particles. Later he performed his celebrated

---

* Invisible ink was known to the ancients. Heating a message scrawled in lemon juice reveals the secrets, as most children know. Urine also works. Thompson used gallotannic acid from powdered nutgalls and developed the message in a solution of iron sulfate. Infusion of nutgalls was considered a remedy for "putrid bilious disorders" (diarrhea), which gave Thompson an alibi for possessing the ingredients.

cannon-boring experiments. Because he was in charge of the munitions works, it was a simple matter for Thompson to borrower a few cannons. With a horse-powered mill, he showed that the heat produced by drilling the cannon was effectively inexhaustible. If heat were a fluid, presumably it would run out sooner or later. Yet "it would be difficult to describe the surprise and astonishment expressed in the countenances of the by-standers, on seeing so large a quantity of cold water heated, and actually made to boil, without any fire." Thompson did point out that the proce-dure was not a good way to boil water for "cooking victuals" and that burning the horses' fodder would be more economical. Thompson's paper was read before the Royal Society in London and caused a sensation.

But the count's career was far from over. He left Bavaria during the war with Austria, returned to England, and almost set off for America upon receiving an invitation to join the United States Military Academy, then being established. Instead he founded the Royal Institution. The Royal Institution, in contrast to the Royal Society, was to be a public institution devoted to the dissemination of scientific and practical knowl-edge. As we have seen in previous tales, names such as Thomas Young, Humphrey Davy, and Michael Faraday became associated with it, not to mention John Tyndall, who shall turn up presently, and the Royal Institu-tion evolved into one of the world's great scientific establishments.

Truth is stranger. Always unsatisfied with his situation and striving to evolve to a higher level, the now world-famous philosopher emigrated to France, where he was entranced by the ladies and married Lavoisier's widow. The match was perfect: they hated each other. Their quarrels "were carried out with such a degree of violence that the . . . Chief Inves-tigator for the State was obliged to take notice of them." Thompson called her "a female dragon." For her part, Mme Lavoisier attempted to get Napoléon to declare her husband persona non grata at court. They sepa-rated in 1809 after four years of bliss, and the count died five years later. It is a wonder that HBO has never made a movie about Thompson.

It is equally strange that Puccini never lit upon André-Marie Ampère as the subject or an opera. Although Ampère is usually thought of in con-nection with physics (and in particular with the unit of electrical current named after him), he made contributions to chemistry as well. He led an exceptionally tragic life. Ampère's childhood in Lyon was idyllic, but in 1792, when he was seventeen, his elder sister died at the age of twenty. Before the end of the year his father, a merchant who had taken up civil duties during the French Revolution, was guillotined. For over a year

Ampère found himself paralyzed with grief. After a long, determined court-ship, he married Catherine-Antoinette Carron, whom he loved deeply, but less than four years later she died of an undiagnosed stomach illness. He married again, to one Jenny Potot, a woman who so detested him that she would not allow him to live in the same house. He learned only by mes-senger of the birth of his daughter, Albine, but his wife was so opposed to the idea of children that she gave Albine up to foster care. Ampère's son by his first marriage, Jean-Jacques, wasted most of his adult life pursuing the famous courtesan Mme Recamier, and Albine later married an alco-holic gambler who held pistols to her head to increase her bravery. After their separation the fellow was packed off to a relative but returned a few years later to live with her in Ampère's house. He was finally arrested when, after threatening Ampère one night with a drawn sword, he ran into the street dressed in his nightshirt, only to be seized by police. Albine herself became increasingly deranged, claiming encounters with the devil, and died a madwoman. Thankfully, Ampère had himself passed away five years earlier, in 1836. He had spent the last decade of his life in poverty. To give you an idea of how such stories go, I had always thought his gravestone read "Peace at last," but the inscription on his final resting place reads, in part, "He loved man / He was simple good and great."

We can forgive Ampère if he was often a morbidly depressed man. And we can admire him, for despite everything he managed to write poetry and made fundamental contributions not only to physics and chemistry but even to mathematics, all without the benefit of a formal education. In chemistry, one of his contributions was to provide a general derivation of Boyle's law (known as Mariotte's law in France), the one that states that the pressure of a gas is inversely proportional to its volume. He submitted this result in 1814. Ampère's derivation, by the way, was made in the con-test of the caloric theory, which shows that the last of the opponents to Rumford's ideas had yet to die off. His derivation was also quite different from the one announced seventy-five years earlier by Daniel Bernoulli. Bernoulli, assuming gases to be made out of invisible particles that obeyed Newtonian mechanics, showed that Boyle's (Mariotte's) law followed. With the further assumption that temperature is proportional to the par-ticle velocity, the perfect gas law, known to all high school chemistry stu-dents, follows immediately. If anyone had paid attention to Bernoulli's work, the notion of atoms might have been established a full century before it in fact was, but Bernoulli's brilliant argument proved to be so far ahead of its time that it was ignored. Another of Ampère's contributions was the announcement of Avogadro's hypothesis independent of Avogadro.

As chemistry students know, Joseph-Louis Gay-Lussac (1778–1850) had found that very nearly two volumes of hydrogen combined with one volume of oxygen to produce two volumes of water vapor—steam. This was very puzzling, for among other reasons, steam is less dense than oxygen—it rises. If steam is composed of oxygen *plus* hydrogen, you would naturally expect it to be heavier than oxygen. In 1811 the Italian Amedeo Avogadro (1776–1856) resolved this paradox with his proposal that equal volumes of gases contain equal number of *molecules.** Avogadro's famous paper in fact is considered the birth of the notion of molecules, but it was ignored for a full half century, thus delaying the acceptance of atoms almost till the opening of the twentieth century (chapters 6 and 7).

One must be a bit careful here. It can't be denied that Avogadro introduced the notion of equal numbers for equal volumes. He says so in his paper, which by the way was also about caloric. But Avogadro's notion of molecules was very different from ours. A decade after Dalton, he does not use the term *atom* but talks about three different sorts of molecules: elementary molecules, integral molecules, and constituent molecules (and even composite molecules); nor does he use these terms in a consistent manner. In speaking of chemical reaction he talks about constituent molecules breaking up into halves and quarters and recombining. I don't believe it is possible to reconcile our view of a molecule with Avogadro's, and he can be regarded as the "father of the molecule" only with slightly Whiggish historical hindsight. One must also ask whether the neglect of his paper is due in part to the obscure way in which it was written. Finally, Avogadro never actually calculated the number of molecules in a standard volume of a gas, the essential "Avogadro's number." That was reserved for Joseph Loschmidt, whom we shall also run into soon.

In any case, Ampère had his own ideas about molecules, Ampère's molecules, like so many other constructs from the minds of natural philosophers, were based on the Platonic solids; they were polyhedra with various numbers of sides and points, depending on the substance. Nevertheless, also inspired by Gay-Lussac's experiments. Ampère realized that the results could be explained if one assumed the equal-numbers-equal-volumes law. His paper, submitted in 1814, suffered the same fate as Avogadro's.

---

* In modern notation, if oxygen were O and steam were HO, and equal volumes of gas contained equal number of molecules, then a volume of steam *would* weigh more than a volume of oxygen. But if oxygen is O_2 and steam is H_2O, then one volume of steam weighs *less* than one volume of oxygen.

*   *   *

The periodic table of the elements would not be complete without its stories. At the Moscow Polytechnical Museum an entire hall is devoted to Dmitri Mendeleev (1834–1907), who is generally regarded as the father of the periodic system and a national hero besides. No mention is made of anyone else. Mendeleev (the last of seventeen children; so much for the firstborn-as-achievers theory) completed his first periodic table on February 17, 1869. Yet it is well known that the German chemist Lothar Meyer independently devised a remarkably similar table the previous year. Unfortunately, Meyer's book was not published until 1870, which resulted in the UBPD (usual bitter priority dispute). It is commonly said that Mendeleev's table was superior because he used it to predict the existence of a several new elements and their compounds. True enough, but in the citation for the Royal Society's Davy Medal, which both men won, the predictions aren't mentioned.

The important thing about the periodic table is that it arrays the elements so that those with similar chemical properties fall into columns. In this, neither Mendeleev nor Meyer was first. Lavoiser's table was merely a list, but by 1817 Johann Döbereiner of Germany pointed out that many elements could be arranged by their properties into groups of threes, which he called triads. Döbereiner's suggestion influenced further work, and by 1862 the French geologist Alexandre-Emile Beguyer de Chancourtois had plotted elements according to their atomic weights (number of protons plus neutrons) on a helix that wrapped around the outside of a tall cylinder. He noticed the elements falling on the same vertical line had remarkably similar properties. This led him to some weird conclusions, for instance that there should be a form of carbon with atomic weight 44. Clearly his results depended on the angle he chose for winding the helix, so there must have been a certain amount of numerology involved. He also failed to include a diagram of the apparatus in his presentation before the French Academy. As we know, a picture is worth a thousand words: no picture, no support.

At the same time the Englishman John Newlands was working on a periodic table. He noticed that elements with atomic weights eight numbers apart seemed to have similar properties. In 1866 he read a paper before the Chemical Society entitled "The law of Octaves and the Causes of Numerical Relations Between Atomic Weights." Newlands must be credited for being the first to propose a periodic law, but his table had other flaws (for instance, it left no room for new elements), and at the meeting one detractor rose to ask him whether he had thought to arrange the elements alphabetically. The editors of the *Journal of the Chemical Society* rejected his paper; only after Mendeleev and Meyer became famous did the Royal Society try to make amends by awarding Newlands the Davy Medal. Mendeleev, by the way, admitted that he had seen Newlands' work. (Mendeleev, on the whole, was a greatly admirable fellow. Like Nikolai Rimsky-Korsakov, he sided with the students during Russia's political upheavals at the dawn of the twentieth century, and like Rimsky-Korsakov, he was dismissed from his post.)

One should also point out that although Mendeleev's table came closest to the modern one, none of the tables were correct. All the nineteenth-century tables arranged elements in order of increasing atomic weight; it was only on the eve of World War I that Henry Moseley determined that the correct sequence of chemical properties was given by basing the periodic table on atomic number (number of protons). He then marched off to Gallipoli.

We have concentrated on chemistry in this introduction. Before turning to the exhibits, we briefly mention a concept from biology, the Gaia hypothesis. The idea that the Earth should be viewed as a living organism is inevitably associated with James Lovelock and Lynn Margulis, who introduced the concept in 1972. At the time, Lovelock and Margulis were unaware that the great progenitor of the Gaia hypothesis had been introduced fifty years earlier by Vladimir Vernadsky. Now, thanks to a new annotated English translation of Vladimir Vernadsky's *Biosphere,* in which Margulis herself seems to have been instrumental, the concept of an integrated terrestrial biological-geological system can be traced back toward its origins. Though entirely unknown in the West, Vernadsky (1863–1945) is one of the most revered Russian scientists. Properly he should be called a mineralogist, but his philosophical writings are extensive, and Russians consider him the founder of geochemistry, radiology, and biogeochemistry. A professor at the University of Moscow, he resigned in 1911 in protest of the tsar's policies and went on to organize and direct a number of laboratories, including the State Radiology Laboratory and what is now called the Vernadsky Institute of Geochemistry and Analytical Chemistry. One of Moscow's largest boulevards is named in his honor.

In his greatest work, *Biosphere* (a term Vernadsky points out was coined by the German geologist Eduard Suess in 1875), Vernadsky attempts to analyze the Earth as a self-contained system. He is particularly concerned with the role of crustal minerals and the interaction of life with the geological and chemical environment. Vernadsky's vision was an all-encompassing one and is still felt today, even by those who don't realize its source. It would be too much to hope that he got everything right. His ideas on evolution, to take one example, seem somewhat strange, even for 1926, but then evolution has always been a sensitive subject, as shall become immediately apparent . . .

17 / The Evolution of Evolution: Erasmus, Charles, Gregor, and Ronald

No scientific theory of the last four hundred years has had as much impact on human thought and culture as Darwin's theory of evolution. Immediately upon the theory's announcement, "survival of the fittest" (a phrase Darwin did not coin) became a nineteenth-century watchword not only of naturalists but of business tycoons and, too often, of racists. But their audacity pales compared to thinkers of our own epoch who extend the concepts of natural selection to social behavior, to psychology, to the propagation of ideas through culture, to the evolution of entire universes. And of course, to this day the legal system is burdened by those who insist that creationism and evolution be accorded equal weight in our school systems; it is part of the never-ending struggle between ecbatology and teleology. The primacy of Darwinism in our scientific and cultural outlook cannot be denied. The only question is, which Darwin?

Even some schoolchildren know that the famously dilatory Charles Darwin was spurred to complete his *Origin of Species* upon receipt of an essay by the naturalist Alfred Russell Wallace containing "exactly the same theory as mine." Judging from recent book covers in which Darwin's revolutionary theory is said to have been sprung on an unsuspecting public in 1859, less well known is that Charles was indebted for many of his ideas to his own grandfather, Erasmus, whom he conspicuously failed to acknowledge.

Called by his principal biographer the most accomplished man of the past three centuries, Erasmus Darwin was born in 1731, the seventh child of Robert and Elizabeth Darwin. Little is known about his parents, except that Robert was responsible for bringing the first known fossilized skeleton of a Jurassic dinosaur (a plesiosaur) to the attention of science, a skeleton that can still be viewed at London's Natural History Museum. Erasmus himself led an outwardly uneventful life, never venturing far from the English Midlands, where he made his home. His adventure was of the mind, and his mind roved far and wide. After studying medicine at Cambridge and Edinburgh, where he is said to have made frequent sacrifices to both Bacchus and Venus, Erasmus moved to Lichfield, England, and became a successful country doctor, the most successful physician in England. He married, had five children by a first wife, Polly, two by a mistress, Mary, after Polly's death, and seven more by a second wife, Elizabeth.

Propagation of the species in action, but such activities hardly exhausted Erasmus' brain. He was an indefagitable inventor, constructing the prototype of a primitive talking machine, a copying machine (pantograph), an artificial bird, a horizontal windmill, and a horse-drawn carriage whose steering mechanism (with independent front and rear axles) is that used by automobiles today. His experiments in electroshock therapy, performed with an instrument of his own devising, stimulated not only the livers of his patients but the imagination of Mary Wollstonecraft and her writing of *Frankenstein*. He was a friend of James Watt, Joseph Priestley, and Benjamin Franklin; was the first to understand cloud formation; made contributions to geology; was an early convert to Lavoisier's theory of combustion; and became, to top it off, the leading English poet of the day, influencing the later English romantics Coleridge, Blake, Wordsworth, and Shelley. He also proposed the theory of evolution.

Poetry and evolution were not unrelated in Darwin's mind. As early as 1770 fossils unearthed during the construction of the Harecastle tunnel convinced Erasmus that life as we know it was descended from a common ancestor. To the family coat of arms, three scallop shells, he even added the motto *E conchis omnia* (everything from shells), but John Seward, a local clergyman, caught sight of it on Darwin's carriage and objected:

> He too renounces his Creator,
> And forms all sense from senseless matter,
> Great wizard he! by magic spells
> Can all things raise from cockle shells . . .
> O Doctor, change thy foolish motto,
> Or keep it for some lady's grotto.
> Else thy poor patients well may quake
> If thou no more canst *mend* than *make*.

Thus stung, Erasmus kept his views to himself for over twenty years. Toward the end of that silence he launched into an active literary career that began with the anonymous publication in 1789 of *The Loves of the Plants,* a fantasy based on Linnean botany about, well, the sex lives of plants:

Sweet blooms GENISTA in the myrtle shade,
And *ten* fond brothers woo the haughty maid.
Two knights before thy fragrant altar bend,
Adored MELISSA! and *two* squires attend.
MAEDIA's soft chains *five* suppliant beaux confess,
And hand in hand the laughing belle address;
Alike to all, she bows with wanton air,
Rolls her dark eye, and waves her golden hair.

Racy stuff this. *The Loves of the Plants* proved an immediate sensation and made Darwin the most famous poet in England. As it turns out, *Loves* was the second half of a long poem entitled *The Botanic Garden,* the first half of which, *The Economy of Vegetation,* was published later, in 1792. *Economy* was even more successful than *Loves,* so overshadowing its former sequel that people forgot that the *Garden* contained two halves. More than a vegetative excursion, *Economy* covers nymphs, gnomes, earth, air, fire and water, and Creation itself:

—"LET THERE BE LIGHT!" proclaim'd the ALMIGHTY LORD,
Astonish'd Chaos heard the potent word;—
Through all his realms the kindling Ether runs,
And the mass starts into a million suns;
Earths round each sun with quick explosions burst,
And second planets issue from the first;
Bend, as they journey with projectile force,
In bright ellipses their reluctant course;
Orbs wheel in orbs, round centres centres roll,
And form, self-balanced, one revolving Whole.

If it initially seems strange that a course in natural philosophy should be ensconced in an epic poem, one should remember that the tradition goes back at least to the ancient Greeks. Indeed, in the sixteenth century, when people still required memory and used it, they memorized mathematical formulas through verse. Only in comparatively recent times has science been scorned by the majority of poets as an unfit subject. And if the above verse holds faint resonances with the modern big-bang theory, the murmurings become louder in one of Darwin's own footnotes:

It may be objected, that if the stars had been projected from a Chaos by explosions, that they must have returned again into it from the known laws of gravitation; this however would not happen, if the whole of Chaos, like grains of gunpowder, was exploded at the same time, and dispersed through infinite space at once, or in quick succession, in every possible direction.

Well, he wasn't a physicist. In another stanza Erasmus also conjectured that the Moon was torn from the Earth, an idea put on a sound scientific footing by his grandson, Sir George Darwin.

The Economy of Vegetation established Erasmus Darwin as the leading English poet of the day, but he was not finished with poetry, and only beginning with evolution. His last long poem, *The Temple of Nature,* published posthumously in 1803, made the evolutionary theme central:

> Organic Life beneath the shoreless waves
> Was born and nurs'd in Ocean's pearly caves;
> First forms minute, unseen by spheric glass,
> Move on the mud, or pierce the watery mass;
> These, as successive generations bloom,
> New powers acquire, and larger limbs assume;
> Whence countless groups of vegetation spring,
> And breathing realms of fin, and feet, and wing.

He is quite explicit on the origin of life:

> Hence without parent by spontaneous birth
> Rise the first specks of animated earth;
> From Nature's womb the plant or insect swims,
> And buds or breathes, with micrscopic limbs.

With the benefit of hindsight, the poem seems extraordinarily prescient—and totally consistent with modern theories. Unfortunately, by the time *The Temple of Nature* appeared, not only was Erasmus Darwin dead but his reputation had plummeted. The decline began in 1794, when he published the first half of his thousand-page tome *Zoonomia*. With this massive treatise Darwin, now "too old and hardened to fear a little abuse," had publicly turned his attention from the poetic classification of vegetables to the prosaic classification of the entire animal kingdom. In chapter 39 of the *Zoonomia,* "Of Generation," Erasmus lays out in detail his conception of evolution. While natural selection may not be *quite* announced in exactly those words, he is unquestionably aware of survival of the fittest. He observes that males of certain bird species are armed with claws the females lack, and concludes these weapons cannot be for fighting external enemies; they are for fighting "for the exclusive possession of the females." He goes on to say, "The final cause of this contest amongst the males seems to be, that the strongest and most active animal should propagate the species, which should thence become improved." He also reiterated his belief "everything from shells," more exactly that all creatures descended from common "filaments," or molecules. In the most celebrated passage of the *Zoonomia* he asks:

> Would it be too bold to image that in the great length of time since the earth began to exist, perhaps millions of ages before the commencement of the history of mankind, would it be too bold to image that all warm-blooded animals have arisen from one living filament, which THE GREAT FIRST CAUSE endued with animality, with the power of acquiring new parts, attended with new propensities, directed by irritations, sensations, volitions, and associations; and thus possessing the faculty of continuing to improve by its own inherent activity, and of delivering down those improvements by generation to its posterity, world without end.

Of course this is not really a question; it is a statement of belief. Darwin's willingness to accept geological time scales for the age of the Earth seems to be incredibly foresighted. In this he was more modern than Charles, who had to contend with Lord Rayleigh and other physicists who refused to accept a geologic age for the Earth. (It goes without saying that here Erasmus outdistances many of our own era.) On the other hand, his idea that species might mutate through "volitions" does seem to veer in the direction of Lamarck, who followed a few years later and believed that characteristics acquired during the lifetime of an organism can be inherited.* Erasmus, though, apparently had more general mechanisms of heredity in mind, understanding that the differentiation of bird beaks "had been gradually produced during many generations by the perpetual endeavor of the creatures to supply the want of food."

To some extent it may indeed have been due to a confusion with Lamarck that Darwin's ideas were discredited. However, this was not the main reason. By the time *Zoonomia* was published, the French Revolution had taken place, and within a few years England would be at war with France. Nobody wanted to hear about the brotherhood of man, not to

---

* Lamarck made many contributions to biology, but he is unfortunately remembered only for his mistaken belief about acquired characteristics. The famous example is that of a giraffe. According to Lamarck, a giraffe stretching its neck to obtain food would pass on the characteristic of an elongated neck to its immediate offspring.

mention a theory that flirted with atheism. And so Erasmus was forgotten. But before he died in 1802 he had passed on his ideas to his son Robert, who in turn passed them on to his son Charles.

Why Charles Darwin, born in 1809, was so reluctant to acknowledge his own grandfather is a matter for psychologists. In his *Autobiography,* Charles overflows with praise of George Lyell, whose *Principles of Geology* so influenced his thinking and which he carried with him on the voyages of the *Beagle.* Among other things he says, "The science of Geology is enormously indebted to Lyell—more so, as I believe, than to any other man who ever lived." As for his grandfather, however, we know that while a student at the University of Edinburgh Charles studied the *Zoonomia* closely. Yet the only thing he has to say about it in the *Autobiography* comes during a conversation with Robert Grant, a lecturer at Edinburgh who tried to convert Charles to the views of Lamarck and Erasmus. "I listened in silent astonishment, and as far as I can judge, without any effect on my mind. I had previously read the *Zoonomia* of my grandfather, in which similar views are maintained, but without producing any effect on me. Nevertheless, it is probable that the hearing rather early in life such views maintained and praised may have favored my upholding them under a different form in my *Origins of Species*. At this time I admired greatly the *Zoonomia;* but on reading it a second time after an interval of ten or fifteen years, I was much disappointed; the proportion of speculation being so large to the facts given."

It all seems a little Freudian. In his defense, Charles and his supporters always took pains to point out that Erasmus was the wild theorizer and Charles the meticulous observer. "I look at a strong tendency to generalise as an entire evil," Charles once wrote to his close friend, botanist Joseph D. Hooker. Indeed, perhaps the main reason *The Origin of Species* occupies its deserved position in the canon of great and influential books is that it overflows with facts and observations. Yet *Origin* also overflows with hundreds of names. Not one of them is Erasmus. Critics did not wait for Freud to interpret this as a significant oversight. Bishop Wilberforce and others pounced immediately. Darwin attempted to make amends by writing an "Historical Sketch" in 1861, but there he mentions Erasmus only in a footnote, and that to dismiss him: "It is curious how largely my grandfather, Dr. Erasmus Darwin, anticipated the erroneous ground of opinion and the views of Lamarck."

Later, at the age of seventy, Charles had a change of heart and decided to write a biography of his grandfather. A guilty conscience? Probably. But the attempt backfired. Charles wrote a short book, allowing it to be published as a 127-page preface to an 86-page essay on Erasmus by Dr. Ernst Krause of Germany. And that touched off an explosion.

Samuel Butler, soon to become author of *Erewhon,* had been an old family friend of Darwin's who fell under the spell of *The Origin of Species* as soon as a copy reached New Zealand. Later Butler discovered Erasmus, Lamarck, and Buffon, whom he preferred. Butler thereupon launched a sustained attack on Charles, claiming that he had filched his grandfather's ideas, and also claiming that as a result of his own proddings Charles deleted some thirty-six references to "my theory" (that is, Charles' theory) from subsequent editions of *Origin.* This seems to be accurate. One of Butler's books, *Evolution Old and New,* was published in 1879, between the publication of Krause's original German essay in the journal *Kosmos* and the appearance of the English edition.

When Butler read the English version he was incensed. Darwin had evidently added a paragraph or two to this "guaranteed" translation of Krause's work, including one that stated, "Erasmus Darwin's system was in itself a most significant first step in the path of knowledge which his grandson has opened up for us, but to wish to revive it at the present day, as has actually been seriously attempted, shows a weakness of thought and a mental anachronism which no one can envy." According to Butler, in fact, "the whole of the last six English pages were spurious matter" and the translation was riddled from beginning to end with additions and deletions. He wrote to Darwin, who replied that Krause himself had made a revised version for the English edition, "so common a practice that it never occurred to me to state that the article had been modified." This is technically accurate, but it was Darwin himself who sent Butler's book to Krause in order that he should make the revision.

If all this weren't enough, Charles' essay as published is a rather dry affair, practically damning Erasmus with faint praise. Of his evolutionary ideas Charles says almost nothing, deferring that task to Krause. It turns out that Charles had allowed his daughter Henrietta, who hated every-thing Erasmus stood for, to censor the work. The final peroration in which Charles praised Erasmus for his generosity and prophetic spirit, and all other favorable passages, were cut.

Butler and Darwin were never reconciled, and in the early editions (now reprinted) of Darwin's *Autobiography* the affair is never mentioned. It was not until 1958 that Darwin's granddaughter, Nora Barlow, restored some cut passages and appended correspondence relating to the controversy. Despite the family's insistence that Charles' reluctance to recognize his grandfather was merely a matter of methodological differences, if there is a wider explanation than the purely Oedipal, it is that, like Einstein, Charles Darwin had a true problem acknowledging his predecessors. Biologists have tended to take the family's word for it, and as a result, evolution more than most theories is almost devoid of a public history. As Darwin put it to T. H. Huxley, "the history of error is unimportant."

Erasable. The anthropologist Loren Eiseley has pointed out that *Origin* was already in proof when Lyell himself caught Darwin in the act of ignoring Lamarck. In the first edition, Darwin omitted "by inadvertence" Wallace's name in the final summary, despite the historic joint announcement and publication of their papers. Well, Darwin told Lyell, he "never got a fact or idea" from Lamarck. One wonders what Charles would have said about Edward Blyth.

In fact he said nothing. Blyth (1810–1873) was a friend of Darwin's and a pioneering naturalist who wrote major papers on heredity and zoology for *The Magazine of Natural History*. In those papers Blyth clearly saw the importance of variation and sexual selection, although he mistakenly interpreted natural selection as a force tending to stabilize species rather than lead to their diversification. For example, in 1835 Blyth wrote, "As in the brute creation, by a wise provision, the typical characters of a species are, in a state of nature, preserved by those individuals chiefly propagating, whose organisation is the most perfect, and which, consequently, by their superior energy and physical powers, are enabled to vanquish and drive away the weak and sickly, so in the human race degeneneration is, in great measure, prevented by the innate and natural preference which is always given to the most comely."

Could Darwin have been unaware of Blyth's work? In the first place, they were friends. In the second place, by tracing Darwin's own footnotes (and even the use of the strange word *inosculate*), Eiseley has been able to determine that Darwin was in possession of and read the same issues of *Natural History* in which Blyth's papers appeared. Darwin later claimed it was Thomas Malthus, who famously argued that population growth must inevitably lead to a human struggle for existence, that gave him the key to natural selection. Yet in *Origin* Darwin repeats many of Blyth's assertions almost verbatim without acknowledgment (he references Blyth only as a taxonomist and observer), and Eisley makes a convincing case that it was Blyth, not Malthus, who started Darwin on the road to natural selection.

If any other precursor to evolution is forgotten more than Edward Blyth, it must be William Lawrence (1783–1867). He is forgotten intentionally because his book *Natural History of Man,* published in 1819, came to conclusions so distasteful to those times (and in part to ours) that it was suppressed. Lawrence, a professor at the Royal College of Surgeons and called by T. H. Huxley "one of the ablest men whom I have known," wrote that:

1. The physical, mental, and moral differences in man are hereditary, and different races are susceptible to different diseases.
2. The different races have arisen through mutations.
3. Sexual selection has improved the beauty of advanced races and governing classes.
4. Characters of races and nations are preserved by the existence of breeding barriers between them.
5. "Selection and exclusions" are the means of change and adaptation.
6. Men can be improved by selection in breeding just as domesticated animal can be improved.
7. The study of man as an animal is the only proper foundation for research in medicine, morals, and even politics.

"The diversification of physical and moral endowments which characterize the various races of man," he wrote, "must be entirely analogous in their nature, causes, and origin, to those which are observed in the rest of the animal kindom and therefore must be explained on the same principles." The origins of man "cannot be settled by an appeal to the Jewish scriptures."

It is almost certainly not the racist overtones of Lawrence's work that so offended English society of the day but the fact that he insisted on treating man as an animal. The Church was outraged, and Lawrence was repudiated by the leaders of his profession, including his own teachers. The lord chancellor refused to allow copyright for the book on the grounds that it contradicted the Scriptures. So crushed was Lawrence by the affair that he withdrew *Natural History* from circulation and clammed up for the rest of his life.

The field was thus clear for Charles Darwin to reap the whirlwind and the glory, but it goes without saying that evolutionary progress did not end with Darwin either. Just six years after the publication of *The Origin of Species* an Augustinian monk, Gregor Mendel, wrote his epoch-making paper "Experiments in Plant Hybridization" and founded the science of genetics. Mendel, born in 1823 into a poor family of Moravian farmers, wanted an education, which he got by joining the Church and becoming a priest. He taught high school as a substitute teacher but in 1850 failed his qualifying exam to become certified for the permanent staff. His monastery then sent him to the University of Vienna for three years, after which he taught physics at the Brünn (Brno) Technical High School until 1868.

Mendel, more enraptured by physics than theology, also had an interest in evolutionary theories, in particular those of Lamarck. During his spare time at the monastery Mendel would notice an atypical plant and wonder how its peculiar traits arose. He began planting typical and atypical varieties side by side and observed that the progeny retained the essential characteristics of the parents, uninfluenced by the environment. Thus was born the notion of heredity. (One wonders what Darwin would have to say about this.) Soon Mendel was crossing pea plants, and after seven years of experimentation he could enumerate the frequency with which various traits (e.g., color, or whether the pea was smooth or wrinkled) were passed from one generation to the next. He found that dominant characteristics (e.g., smooth peas) outnumbered recessive characteristics (wrinkled peas) in the first generation of offspring by three to one, regardless of the particular trait involved. Subsequent generations also displayed repeatable patterns. In 1865 Mendel read his results before the Society for the Study of Natural Science at Brünn, results that would shake the foundations of biology—and he was ignored.

So profound was the silence that in some versions of the tale a discouraged Mendel overthrew his work in favor of administrative duties, to which he devoted himself until his death in 1884, sixteen full years before his results would be simultaneously and dramatically rediscovered by German botanist Carl Correns, Dutch botanist Hugo de Vries, and Austrian agronomist Erich Tschermak von Seysenegg. I do mean simultaneously and dramatically: the three papers were all submitted *to the same journal* within two and one-half months of one another! Correns and de Vries, as it happened, were mainly interested in theoretical concepts, but Tschermak recognized the practical value of the laws of heredity for agronomy, and it was largely through his efforts that genetics became the foundation of modern plant breeding and, by the twenty-first century, much more.

It is the great thirty-four-year silence that captures the imagination of textbook writers, almost certainly because it resonates with the universal sense of neglect shared by scientists everywhere. You can even read about Mendel's neglect in the current edition of *Britannica,* which is a little strange given that *Britannica* itself cited Mendel's work in the ninth edition, from 1881 to 1895. It was this mention that eventually led de Vries to look up Mendel's original paper once he had written his own. A number of other researchers, in particular W. O. Focke, cited Mendel at least fifteen times. Mendel's work also made its way to America and was listed in the *Royal Society Catalogue of Scientific Papers.*

None of this means that Mendel was *understood,* and perhaps this is the main point. Mendel himself corresponded with the great German biologist Carl Wihelm von Haegeli, who persisted in being unable to grasp Mendelism, despite tutelage from Mendel himself. Focke, though he repeatedly cited Mendel, appears, in the great tradition, not to have read him. Also, where today we see genetics as the foundation of evolution, at the time of Mendel and Darwin that concept still lay in the future. It is often said that Mendel regarded inherited traits as immutable and therefore rejected evolution. This in itself is not so obvious; Mendel is clearly aware of the overwhelming relevance of his experiments to evolutionary theory—and says so at the start of his paper. From his marginal notes to *Origin of the Species,* moreover, it is clear that Mendel accepted the idea of natural selection. On the other hand, Darwin's ideas on the fundamental cause of the variability of the species were very confused, even wrong (much more like Lamarck than he would admit), and he never seems to have realized that he held the correct cause of natural variation in his hands, literally: Focke's book had crossed his path, with its citations of Mendel, but whether Darwin ever opened it nobody knows.* The safest thing to say is that the causal connection between hereditary traits and evolution of the species had not yet been established in Mendel's mind, or anyone else's.

Certainly if Mendel was ignored, then a new word needs to be invented to describe what happened to his predecessors. As early as the sixteenth century, naturalists were attempting to account for the varigated colors that could appear on a single ear of corn, and the idea of dominant characteristics on corn was noticed by several amateurs, including Cotton Mather in 1716. The well-known botanist Thomas Knight, in a 1799 letter to the Royal Society, described how he crossed two varieties of pea plants and found both dominant and recessive traits in the first generation, then back-crossed the hybrids with the recessive parents to produce both dominant and recessive types. His results were extended to three generations by John Goss, who discovered three (pre-)Mendelian types in

---

* For example, Darwin believed that each part of the body produced somatic material called "gemmules," which were injected into the bloodstream and carried to the sperm and ova. Environmental factors or direct action on the organism would modify the gemmules, thus transmitting variations to the next generation. He called this hypothesis "pangenesis."

the third generation; Knight himself confirmed these results. Similar studies were carred out on muskmelons in 1826 by Augustin Sagaret. None of these investigators produced numerical ratios for the various traits, but the Silesian priest Johann Dzierzon (almost) did for bees in 1854.

Dzierzon (1811–1906) is considered one of the seminal names in bee breeding. In an 1854 issue of *Der Bienenfreund aus Schlesien* Dzierzon published a paper on bee hybrids and noted that if the queen bee is herself a hybrid of two types (German black bees and Italian gold bees), she will produce not intermediate offspring but equal numbers of drones of the two types, "as if it were difficult for nature to fuse both species into a middle race." And although he doesn't quite spell it out, the implication is that if black is dominant and gold is recessive and the queen harbors both, then under random fertilization from drones you should get three black bees for one gold bee.

There is some indirect evidence that Mendel built on the work of his predecessors. One of the few works he cites in his own investigations is that of a German named C. F. von Gärtner. In 1849 Gärtner published a book that was a detailed summary and explanation of the earlier research in hybridization. Mendel also had an interest in bees and kept careful notes about their colors, tendency to sting, and sweeter aspects. Mendel's chief biographer, Hugo Iltis, feels he quite probably knew of Dzierzon's work, which was hotly debated at the time.

By all accounts Mendel was a sincere and beloved figure, and there is no suggestion of him swiping anything from his forebears, who were then interred by his own great work and the harsh wash of history. In any case, it can't be denied that the Augustinian's 1865 paper was a model of experimental clarity in both execution and exposition. Indeed, Mendel was perhaps a little too clear for his own good, as seventy years later he came in for some interesting criticism by Ronald A. Fisher.

Sir Ronald Fisher (1890–1962) may be the most influential person of the twentieth century whose name is totally unknown to the general public. A professor at University College, London, and then at Cambridge, he was largely responsible for putting the science of statistics on a modern, mathematical footing. Fisher pioneered the use of randomization in experimental design, as well as the use of significance tests, and he derived many common statistical distributions. The Fisher t-test and the F distribution are both named in his honor. A quirky, argumentative fellow who did not suffer fools gladly, by avocation Fisher was a geneticist—some fans say the greatest of all time—and it was largely through his efforts that biologists began to understand that genetics was at the basis of evolution, not incompatible with it. In 1936 Fisher wrote a paper in which he analyzed

Mendel's experiments and concluded that Mendel's numbers were per-fect—a little too perfect. By the usual rules of statistical sampling, the odds that Mendel could get the results he did were about one in thirty thousand. One detects the slightest grin as Fisher sums up his conclu-sions: "Although no explanation can be expected to be satisfactory, it remains a possibility among others that Mendel was deceived by some assistant who knew too well what was expected. This possibility is sup-ported by independent evidence that the data of most, if not all, of the experiments have been falsified as to agree closely with Mendel's expecta-tions." Well, Newton also fudged his data. But as we know, statistics can't tell you what occurred, only what is likely to have occurred; maybe Mendel was lucky. Maybe we were lucky too. If Mendel's numbers had been way off, the science of genetics would never have been invented, at least until 1900.

18 / Dreams with Open Eyes: Kekulé, Benzene, and Loschmidt

"There are two gates of Sleep," Virgil, paraphrasing Homer, tells us. "One is of horn, easy of passage for the shades of truth; the other, of gleaming white ivory, permits false dreams to ascend to the upper air." That the poets grant ascent of false dreams by way of ivory while truth passes through lowly horn surely contains a message. But like the dreams themselves, the missives sent by poets—and chemists—are often not intended for mortals to decode.

There can be no more famous dreams in the history of science, and no dreams of greater consequence in all history, barring perhaps Constantine's, than those that came to August Kekulé in the years around 1858 and 1862. The first took place one summer evening when he fell into a reverie on a London omnibus. Atoms began to combine into groups of two, three, and four. Larger atoms attracted smaller ones; entire chains formed and attached smaller atoms to their ends. Suddenly the conductor shouted, "Clapham Road!" Kekulé awoke and spent the rest of the night sketching his dream onto paper. The result was the theory of organic molecular structure, which revolutionized chemistry.

Kekulé's second dream is even more famous. As he tells it:

> Something similar happened with the benzene theory. During my stay in Ghent, I lived in elegant bachelor quarters in the main thoroughfare. My study, however, faced a narrow side-alley and no daylight penetrated it. . . . I was sitting writing on my textbook, but the work did not progress; my thoughts were elsewhere. I turned my chair to the fire and dozed. Again the atoms were gamboling before my eyes. This time the smaller groups kept modestly in the background. My mental eye, rendered more acute by repeated visions of the kind, could now distinguish larger structures of mainfold conformation: long rows sometimes more closely fitted together all twining and twisting in snake-like motion. But look! What was that? One of the snakes had seized hold of its own tail, and the form whirled mockingly before my eyes. As if by a flash of lightning I awoke; and this time also I spent the rest of the night in working out the consequences of the hypothesis.

The result of Kekulé's second dream was the structure of benzene. To give an idea of that discovery's importance, aspirin is the simplest benzene derivative. Most medicines are aromatic, meaning based on benzene. The entire dyeing industry, whether radioactivity pioneer Abel Niepce de Saint-Victor knew it or not, is based on benzene. The structure of benzene is one of the cornerstones of modern chemistry, and of modern life.

It is also a cornerstone of both New Age psychology and the history of science, if one can accept that paradox. A Google Web search on "Kekulé dream" reveals approximately eight hundred hits. A large number are New Age sites that concern the importance of dreaming, of the unconscious mind. But the vision of Kekulé dozing off in front of his fireplace with the serpent Ouroboros spinning above him has also become one of the most powerful in the lore of science. I remember from decades back seeing a cartoon of the scene in a children's encyclopedia, and more recently essentially the same cartoon appears in Royston Roberts' book *Serendipity*. Scientists, no less than New Agers, appeal to Kekulé's dreams to illustrate the importance of nonlinear thinking, the need to reach an impasse, to retreat and enter an altered state. It is serendipity without serendipity; the mind has been preparing itself and must be released. Any working scientist can avow to this necessity. Any working scientist—any creative person—can tell of going to sleep at night with an unsolved problem nagging at the mind, only to wake up in the morning with it solved. Henri Poincaré, in another famous story along these lines, tells of how the theory of Fuchsian functions came to him one day while boarding a streetcar, after months of apparently fruitless hammering.

The critical role of the unconscious in guiding creativity can hardly be denied, whether it is for chemists dreaming of snakes or poets dreaming of the pleasure domes of Kublai Khan. What is interesting about the vast Internet trove on Kekulé is that virtually none of the sites so much as question the veracity of his account. Even a supposedly serious book, such as *Serendipity*, like most popular scientific histories, makes no real attempt to sort dreams from life. Such is the hold Kekulé's dreams have on the collective unconscious. But in this case, it might have been Lobachevsky, reincarnated as Tom Lehrer, who had it right in the famous song: "Don't shade your eyes, plagiarize."

Freidrich August Kekulé, later ennobled as von Stradonitz, was born in 1829 in Darmstadt, Germany, and initially intended to study architecture at the University of Giessen. A chemistry teacher, however, turned his life in another direction, and after graduating from Giessen, Kekulé went on to study in Paris and England, then to teach at Heidelburg and Ghent, where he remained from 1858 until 1865. In that year he took up a chaired professorship in Bonn, a position he held until his death in 1896. Although he was apparently not much of a practical chemist, his standing as a theorist became enormous, and he is remembered as one of chemistry's great teachers; three of the first five Nobel prizes in chemistry went to his students.

Kekulé's immortality rests on the discoveries that resulted from the two dreams, the theory of organic molecular structure in 1858 and the structure of benzene in 1865. At the core of the structural theory is the idea that carbon is quadrivalent—can bind to four other atoms. Kekulé seems to have had this idea in daylight, somewhat before the first dream: He first mentions it casually, without elaboration, in the footnote to a paper of 1857. Carbon's quadrivalence allows nature to construct long chains of carbon atoms surrounded by hydrogens—hydrocarbons such as ethanol and gasoline—and myriad other organic molecules. This is what came to him during that fateful reverie on the London omnibus—organic structure. His famous paper that solidified the dream, "On the Constitution and the Metamorphoses of Chemical Compounds," appeared in 1858 in Liebig's *Annalen der Chemie.*

Before proceding, one should highlight the obvious: Kekulé reported neither of the celebrated dreams at the time they allegedly took place. The first mention of them was in his famous speech at Berlin City Hall in 1890 on the twenty-fifth anniversary of the announcement of benzene's structure. The above excerpt is taken from that speech. Twenty-five or thirty-two years is quite sufficient for memory to rewire itself, though in the case of the first dream, there is not exactly hard evidence that it took place.

There *is* hard evidence that Kekulé was not alone in proposing his seminal concepts. According to C. A. Russell in his *History of Valency,* the idea of quadrivalence was first introduced in 1855 by William Odling in a lecture at the Royal Institution, undoubtedly in the same ampitheater where Davy demonstated his lamp, and where Young did not describe his immortal experiment. Odling was apparently not interested in priority disputes and never took on Kekulé.

More well known these days is that a young Scottish chemist, Archibald Scott Couper (1831–1892), independently hit on both quadrivalence and organic structure at virtually the same instant as Kekulé. Scott, who was then studying in Paris under one of Kekulé's teachers, Charles-Adolphe Wurtz, submitted a paper on the subject to the French Academy, but Wurtz sat on it and publication was delayed. Three and a half weeks before his paper was read, Kekulé's appeared.

If Kekulé had not known of Couper's work beforehand, he did shortly thereafter and wrote a note for the proceedings of the French Academy attacking Couper. "In two memoirs which have appeared in Liebig's *Annalen* . . . I have put forward different views, which, in my opinion, should furnish a clearer insight into the constitution of chemical compounds." Kekulé then takes credit for the property basicity (the opposite of acidity) and declares, "If Mr. Couper thinks he has discovered the cause of this difference of basicity in the existence of a special kind of affinity, I am the first to admit that I have no right to contest the priority in this."

In less technical language, Kekulé refuses to admit to anything. As a result of the delay in announcing his work, Couper felt betrayed and became extremely abusive toward Wurtz, who then expelled him. Couper returned to Scotland, becoming a laboratory assistant at the University of Edinburgh. Later he suffered a nervous breakdown and never recovered fully. "He was a complete wreck," a colleague wrote. Couper spent the remainder of his life in Kirkintilloch, near Glasgow, cared for by his mother and dying in obscurity. In his 1858 papers, by the way, Couper also introduced the modern chemical shorthand of representing chemical bonds by dashes: H–O–H.

Couper had slipped entirely into oblivion until the opening of the twentieth century. His fate was uncovered by an eminent student of Kekulé, Richard Anschütz, who succeeded him at Bonn. In 1885 Anschütz was trying to reproduce Couper's work on a certain reaction involving salicylic acid. Kekulé had himself tried some years earlier, failed, and dismissed it as possibly fraudulent, but Anschütz followed Couper's methods more carefully and succeeded where the master had not. His interest sparked, Anschütz and his prominent colleague Alexander Crum Brown were able to piece together what we know today about the unfortunate Scotsman.

Anschütz did not stop with Couper. A few years later he was reading— presumably rereading—Kekulé's famous 1865 paper on the structure of benzene. He came across footnote 2, in which Kekulé states that his structure is "preferable to the modifications proposed by MM. Loschmidt and Crum Brown." Modifications proposed by Loschmidt? Crum Brown was famous, but Loschmidt? Anschütz set off on some arduous detective work and eventually discovered that in 1861 Joseph Loschmidt (1821–1895), an unknown high school teacher in Vienna, had published a book entitled *Chemische Studien,* which contained 368 chemical structures. Anschütz's astonishment is easy to imagine. The circular structure of benzene is there. So too are over one hundred other aromatic compounds (some of which were unimagined at the time) and the structure of ozone, as well as many modern notational conventions.

Anschütz was third a disciple of Kekulé, second a scientist, and first a human being. He felt duty-bound to republish Loschmidt's work, which he did in 1913. Nevertheless, Anschütz was initially convinced that Kekulé could not have known of this obscure work. Later he had reason to change his mind. At work on a biography of Kekulé, Anschütz discovered a letter from his teacher to Richard Ehrlenmeyer (of the flask) dated January 4, 1862, just months after Loschmidt's book was published. In it Kekulé ridicules the high school teacher's structures as "*Confusions formeln*" [*sic*] Kekulé had seen Loschmidt's book. Not only had Kekulé seen it, he had read it.

Treachery or oversight? Well, as they say, dreams do not require footnotes or literature citations. One can play armchair psychologist and invent any number of scenarios to explain why Kekulé did not credit Loschmidt: he forgot, intentionally (a recurring exhibit at the Panopticon) or unintentionally; the footnote he did insert into his benzene paper was the Freudian betrayal of a guilty conscience; his treatment of Couper shows he was not above such things. To be sure, since the nineteenth century enemies and detractors have accused Kekulé of ignoring and appropriating the work of others, in particular the work of the Frenchman Auguste Laurent and the Englishman Sir Edward Frankland.* Even the famous Russian composer Aleksandr Borodin—more renowned as a chemist during his lifetime—wandered into the fray. Borodin wrote to his wife, undoubtedly between acts of *Prince Igor,* that Kekulé reproached him for appropriating a certain idea, forcing Borodin to join battle against him in print. Then there is Butlerov . . .

But the dream . . . Well, some reasonably persuasive evidence has emerged that Kekulé fabricated the whole thing. Kekulé did not recall the dreams until 1890, in his Berlin speech. The newspaper accounts differ greatly from the final published version—only three out of twenty-eight reporters mention dreams, and nobody mentions snakes. It turns out no stenographer was present to record the event, and Kekulé preferred to have

---

* Auguste Laurent used an early hexagonal representation of benzene compounds. Sir Edward Frankland brought the concept of valence into the open before Kekulé.

"his rubbish of a speech burned and not allow anything to be printed." He rewrote it completely three weeks later. What's more, he alleged to have had the dream in Ghent in 1861 or 1862 but continued for another three years to portray benzene in his "sausage" notation—which is nothing like snakes biting their tails. Most enticingly, here is precisely what Kekulé says about that immortal reverie on the London omnibus:

> I fell into a reverie, and lo, the atoms were gamboling before my eyes. Whenever, hitherto, these diminutive beings had appeared to me, they had always been in motion; but up to that time I had never been able to discuss the nature of their motion. Now, however, I saw how, frequently, two smaller atoms united to form a pair; how a larger one embraced the two smaller ones; how still larger ones kept hold of three or even four of the smaller; whilst the chain, dragging the smaller ones after them but only at the ends of the chain. I saw what our past master, Kopp, my highly honored teacher and friend, has depicted with such charm in his *Molecular-World*; but I saw it long before him.

"Clapham Road!" Kekulé wakes up.

The important thing here is the last sentence, which is sometimes excised in translation (for instance, in Royston Robert's *Serendipity*). In his book *Molecular-World*, published in 1881, Hermann Kopp wrote about looking through a window into a club for molecules:

> Purpose of the club: dancing. We are not admitted: they are very exclusive here. But we may watch through the window panes. There they are dancing, the molecules! Now faster, now less fast, depending on whether they are heavier or lighter, they pass before our curious eyes. There they are dancing, the molecules, each as a whole, but within each molecule also each atomic group . . . has its extra amusement dance.

Kekulé in 1890 remembering a dream of 1858? Or Kekulé in 1890 acknowledging a book he had read in 1881?

What of Loschmidt, the unknown Viennese high school teacher who had hit on the structure of benzene four years before Kékule? In recent years, Anschütz's discovery of Loschmidt has been resurrected by Alfred Bader, the founder and ex-president of Aldrich Chemical, whose book *Adventures of a Chemist Collector* provided the stimulus for this account. Since 1995 Bader and colleagues have been campaigning for recognition of Loschmidt as the father of molecular modeling. The heat generated by their claims that Kekulé quite possibly plagiarized Loschmidt's formulas has, not surprisingly, been considerable. Critics have replied that Loschmidt's formulae weren't right, that his diagrams weren't meant to portray the actual positions of atoms, and so forth—criticisms that apply more aptly to Kekulé himself. The creak of falling idols is loud, as are the shouts of those being crushed beneath them. The real fear is, are we to replace the great Kekulé by an unknown high school teacher?

To physicists, actually, Joseph Loschmidt is not unknown. Born the son of a poor Bohemian farmer, Loschmidt graduated from the Polytechnic Institute in Vienna in 1846 but failed to find an academic position. Like Davy Crockett before him, he considered emigrating to Texas but ended up wandering through central Europe, first working at a paper factory, then starting a business to make potassium nitrate—the crucial ingredient in gunpowder—for the Austrian government. The business failed because the government had fixed prices that failed to take account of inflation during the Austro-Hungarian war. In the early 1850s Loschmidt returned to Vienna, penniless, and took up a position as a concierge.

Indomitable in spirit, Loschmidt eventually qualified himself as a secondary school teacher and in 1856 began teaching high school in Vienna. Five years later, at the age of forty, he published at his own expense the *Chemische Studien,* his first scientific work. It was ignored by the establishment (except apparently by Kelulé); now, nearly 150 years afterward, it has been called the chemical "monograph of the century."

Loschmidt then turned his energies to the direction of physical chemistry. In 1866 he became the first person to calculate the diameter of a molecule and, using it, became the first person to find a value for Avogadro's number (or Ampère's). As any high school chemistry student knows, Avogadro's number is the number of molecules in a mole of a gas, that is, in 22.4 liters. Avogadro actually never found a value for the number (see the introduction to the Domain of Chemistry and Biology), and it was Loschmidt who managed the first decent calculation, arriving at a figure that is about ten times less than the presently accepted value. To be sure, his method was very similar to the one Einstein would use about forty years later (chapter 8), and to this day in German-speaking countries Avogadro's number is referred to as "Loschmidt's number."

Loschmidt's brilliance was recognized by others, in particular by the famous physicists Josef Stefan and Ludwig Boltzmann, who befriended him. At Stefan's urging, Loschmidt was finally granted a faculty position at the University of Vienna in 1866, where he became a full professor in 1872. It was there that he raised his famous and cogent objections to his friend Boltzmann's celebrated H-theorem. To go into this monumental controversy would take us far afield, but the greatest mystery in physics—then and now—is why time seems to go forward, even though the fundamental equations of physics don't care which way time proceeds. The chief phenomenon by which we perceive the unidirectional flow of time is the increase of entropy, that nearly mystical quantity, which since Boltzmann, in fact, has been popularly believed to be the measure of disorder and randomness. Boltzmann thought he could derive entropy's increase—enshrined in the infamous second law of thermodynamics—from fundamental Newtonian physics, thereby resolving the great paradox. But Loschmidt showed that Boltzmann's proof subtly assumed its own conclusion, which spurred Boltzmann on to more elaborate versions of his theorem. Yet it is not clear to this day that Loschmidt's objections have been adequately answered.

With such achievements, how is it that Loschmidt's chemical work found the Oblivion File? And should the mantle of "father of molecular modeling" be transferred from Kekulé to Loschmidt? As to the first, Loschmidt's fate was partly of his own devising. By all accounts a shy and self-effacing man, he had no knowledge of the Zel'dovich principle, "Without publicity there is no prosperity." There was of course the bad luck of his beginnings, his random walk in life. All of which shows the great arbitrariness of fame and destiny.

As for Kekulé: In Thoreau's words, "dreams are the touchstones of our characters," and Bader and allies indeed argue that the mantle should be transferred. If Kekulé had acknowledged Loschmidt, molecular modeling would have been with us a century sooner than it was. The bitter fact is that Loschmidt's formulae had no impact on the development of chemistry, and so in this world he was simply not the father of molecular modeling. In any case, Loschmidt himself, like Bach, did not spring forth from a field of silence. The lesson of Couper, Loschmidt, and Kekulé is the same lesson we have encountered at every corner in the Panopticon: that history is infinitely more organic than a wall chart plastered with two dozen names and three-line captions.

19 / Chance, Good and Bad: Penicillin

It is in all likelihood the most famous accident in the history of science, eclipsing even Henri Becquerel's discovery of radioactivity. Alexander Fleming, a quiet Scottish bacteriologist working in a London hospital, returns from a two-week holiday to his laboratory. Sorting through the plates of bacterial cultures he left on his lab bench during his vacation, he notices that one has been contaminated by an unwanted mold. About to throw out the dish, he spies a clear ring around the mold where the bacteria has disappeared. "That's funny," he remarks to his assistant. The rest is history.

The discovery of penicillin was not merely a great accident; it must be seen as a pivotal event in human history. For the great war over the past one hundred years has been not World War I, or World War II, the war on drugs, or the war on terrorism. It has been the war against disease. Only with the opening of the twentieth century did military might approach the efficacy of typhus in killing soldiers, and during the nineteenth and twentieth centuries, tuberculosis alone was responsible for the deaths of a thousand million human beings. The innocent mold in Fleming's petri dish—not too different from the mold found on ordinary blue cheese—turned out to be the greatest wonder drug of its time and forever altered the treatment of bacterial diseases. The number of lives saved by penicillin can only be reckoned in the millions, if not hundreds or thousands of millions. The story of its discovery was indeed history.

With gaps. What actually happened on that celebrated day in 1928 will forever remain a bit mysterious, because at the time nobody thought it important enough to pay much attention, and all the accounts are reconstructed through the haze of fifteen years. By the time of his momentous discovery, Fleming, born in 1881, was a recognized bacteriologist. Six years earlier he had discovered that his nasal mucus and tears contained an antibacterial agent, which he dubbed lysozyme. Lysozyme proved ineffective against the more deadly microbes, but in pursuing the action of natural antiseptics against bacteria, Fleming cultivated hundreds of bacteria colonies. It was in one of those petri dishes that on the morning of September 3, 1928, penicillin suddenly appeared. Well, September anyway. Where the mold came from has never been exactly determined, but it probably originated in a downstairs lab and blew in through Fleming's open door. Neither did Fleming, apparently, notice the mold before throwing out the immortal petri dish. It was sitting in a tray of Lysol, having miraculously escaped disinfection, when Fleming casually picked it up and showed it to his assistant, who had stopped by. On this second go-round Fleming noticed the clear ring around the penicillin where the staphylococci had been.

In any case, something along these lines did happen in early September 1928, and Fleming spent the rest of the day showing the dish to his colleagues, who met his discovery with a shrug. Fleming himself, though, was perspicacious enough to realize he had found something unusual, and preserved the dish, which can today be viewed in the British Museum. He was also dogged enough to persist, and set two graduate students the task of producing penicillin for experiments. And early on he recognized penicillin's potential as an important medicine.

All this really happened. What also happened is that Fleming became enormously famous. Nowadays it is difficult to comprehend the magnitude of his celebrity: dozens of honorary degrees, including from Harvard and London, hundreds of awards and decorations, keys to the city, audiences with the pope, fellowship in the Royal Society, a knighthood, the Nobel prize for medicine, upon his death an enormous funeral procession, burial in St. Paul's Cathedral, a crater on the moon named in his honor.

What happened as well was that, as Fleming's fame skyrocketed in the 1940s when penicillin finally became available, the legend of the discovery grew along with it. We are privileged in this case to be able to glimpse a little of the legend's creation. Official biographer André Maurois, writing in the late 1950s after Fleming's death, tells the tale as it is commonly told: Fleming alone had the foresight to see the application of his research and spent years struggling to convince the powers that be to act. "There is something deeply moving in the spectacle of this shy man with his burning faith in the capital importance of a piece of research trying, in vain, to persuade those who alone could have made its practical application possible, to see as he did." That, as will become clear, did not happen. Facts, of course, should never get in the way of a good story, and in Maurois' hands the hero has become not only a bacteriologist but a biblical scholar. Here is Maurois' account of the immediate aftermath of the discovery, slightly condensed:

> Fleming initially showed the mold to a mycologist, C. J. La Touche, who decided it was *penicillium rubrum*. . . . Two years later the famous American mycologist Thom identified the fungus as a *penicillium notatum*. La Touche apologized to Fleming for having misled him. Fleming learned from Thom's book that the *penicillium notatum* had been originally recognized by a Swedish chemist, Westling, on a specimen of decayed hyssop.

This reminded Fleming of the Covenanter, of Psalm 51: "Purge me with hyssop and I shall be clean"—the first known reference to penicillin.

Yet Maurois was far from alone as a creator of icons. Undoubtedly fired by his predecessor's romantic vision, the teleplay writer Ian Curteis decided to go one better for a BBC broadcast in 1972. This conversation between Fleming and his lab director, Sir Almroth Wright, is meant to take place at the moment of discovery in 1928 and makes both of them into biblical scholars:

WRIGHT: What is the stuff, anyway?

FLEMING: I think it's a *Penicillium—Penicillium notatum* or something like that. First identified by Westling on a piece of the herb, hyssop.

WRIGHT: Purge me with hyssop and I shall be clean . . .

FLEMING: Psalms.

WRIGHT: Fifty-three.

FLEMING: Fifty-one.

WRIGHT: Fifty-three.

(pause)

FLEMING: Verse seven.

No wonder the public has a warped view of science.

Thus: Fleming discovered penicillin, recognized its importance, single-handedly campaigned for its production, and bestowed upon the world a miracle. What did not happen, by Maurois' account and those of others, was sixty years of what went before and fifteen years of what came after. By my count, Fleming was at least the sixth person to discover the action of penicillin. A book by Papacostas and Gaté, *Les Associations microbiennes,* published just before Fleming made his discovery, is devoted to the extensive literature on the antagonism between molds and bacteria. Some early investigators used methods almost identical to Fleming's, and others went farther than he did by injecting live animals with mixtures of mold and bacteria.

Apart from the cases published by Papacostas and Gaté, in 1871 Sir John Burdon-Sanderson, who had worked in the very same hospital as Fleming would, wrote an extensive report to the Privy Council on the "development of microzymes." Those were the days when the possibility of spontaneous generation was still taken seriously and when most naturalists put bacteria in the animal kingdom. Nevertheless, Sanderson did detail how mold of the *Penicillium* group inhibited growth of these micro-zymes in cultures. Sanderson's observation stimulated none other than Joseph Lister to study the matter himself. Then engaged in his search for a nontoxic antiseptic, he confirmed Burdon-Sanderson's results and in 1872 wrote to his brother that he would like to test whether *Penicillium glaucum* would inhibit the growth of bacteria in human tissue. Ten years later he was presented with the dramatic opportunity for a test. A young nurse who had been seriously injured in a street accident had sustained a wound that refused to heal under a variety of antiseptics. Eventually a medicine was tried that worked miraculously: *Penicillium.*

In a paper read before the Royal Society in 1874, William Roberts, who was following up Sanderson's investigations, mentioned in a footnote that "I have repeatedly observed that liquids in which the *Penicillium glaucum* was growing luxuriantly could be with difficulty be artificially infected with bacteria." On the other hand, Roberts conceded, "the *Penicillium glaucum* seldom grows vigorously, if at all, in liquids which are full of bacteria."

Less equivocal and certainly more extensive were the investigations into penicillin undertaken by John Tyndall, who announced his findings before the Royal Society in 1876. Tyndall, Ireland's greatest scientist, deserves to be better known. Born in Leighlinbridge in 1820, as a youth he surveyed prospective railway lines for the government. Moving to England he eventually found employment as a mathematics instructor at a secondary school, then at the ripe age of twenty-eight went on to Germany, where, knowing virtually no German and scarcely more mathematics, he managed to get a doctorate in physics in two years. Failing to secure a better position than his instructorship back in England, he lived several years hand to mouth. But in 1853 fate smiled. After an evening lecture at the Royal Institution, Michael Faraday himself shook his hand. Within several months he was elected professor of natural philosophy there, and he remained associated with the institution for the next thirty-four years, succeeding Faraday as superintendent of the house. Tyndall was a renowned publc speaker, jumped unhesitatingly into combat, and for his early defense of Darwinism was denounced as "a materialist and an atheist." One of the most noted alpine mountaineers of his day, Tyndall eventually died in 1893 when his wife gave him an overdose of chloral. The *Oxford English Dictionary*, perhaps incorrectly, credits him with the first modern usage of the word *physics,* dating from 1860: "M. Agassiz is a naturalist and appears to have devoted but little attention to the study of physics."*

* William Whewell coined the terms *scientist* and *physicist* in 1840.

To be sure, today Tyndall would probably be called a physicist, and he is best remembered for his pioneering investigations of the absorptive properties of atmospheric gases. He seems, in fact, to have been the first person to predict the greenhouse effect, which he did in 1861:

> On a fair November day the aqueous vapour in the atmosphere produced fifteen times the absorption of the true air of the atmosphere. It is on rays emanating from a source of comparatively low temperature that this great absorptive energy is exerted; hence the aqueous vapour of the atmosphere must act powerfully in intercepting terrestrial radiation; its changes in quantity would produce corresponding changes of climate. Subsequent researches must decide whether this *vera causa* is competent to account for the climatal changes which geologic researches reveal.

Strangely, Tyndall's preoccupation with the atmosphere indirectly led to his work on penicillin. In the great meandering way of science, Tyndall realized that London air, "which is always thick with motes," caused light beams passed through it to scatter in all directions, whereas "pure air" allowed light to pass unimpeded. Perhaps, he wondered, dirty air not only scattered light but was also responsible—somehow—for the spontaneous generation of microorganisms. Tyndall built an apparatus to test the hypothesis—a sealed box through which he could pass a light beam and watch it scatter. After letting the box stand for a few days, any floating material would be collected on the walls, which he had made sticky with glycerine; the light no longer scattered. The air was pure. Within the chamber also stood a series of test tubes; via a pipette that penetrated the chamber through a rubber seal, Tyndall could introduce a variety of liquids—water, tea, coffee, urine, infusion of grouse. The idea was to compare what happened in these test tubes, exposed only to "moteless air," with a set of identical tubes left out in the dirty laboratory.

In almost all cases, Tyndall found that the liquids within his test chamber remained clear, but that within a few days those out in the lab were covered with *Penicillium*. By 1876 he was ready to report to the Royal Society:

> The *Penicillium* was exquisitely beautiful. Its prevalent form was a circular path made up of alternate zones of light and deep green. In some cases the liquid was covered by a single large patch; in others there were three or four patches, each made up of differently coloured zones. . . . Three kinds of *Penicillium* seemed struggling for existence. . . . In every case where the mould was thick and coherent the bacteria died, or became dormant, and fell to the bottom as a sediment. . . . The growth of

the mould and its effect on the bacteria are very capricious. . . . In some tubes the former were triumphant; in other tubes of the same infusion the latter was triumphant.

It is clear from this extract that Tyndall, just as Burdon-Sanderson and Roberts before him, realized the mold impeded the growth of bacteria. On the basis of this paper Web denizens often claim for Tyndall the discovery of penicillin. Not so. Throughout his address to the Royal Society, Tyndall refers to Sanderson's experiments, though he is admittedly a little vague as to precisely which ones, but from the context there can be little doubt that he is talking about Sanderson's report to the Privy Council. Tyndall even seems to have collaborated with Sanderson, providing him with pure water for experiments. Tyndall, on the other hand, doesn't mention Roberts; evidently they were both borne to their conclusions by the moted air of science, which carries all contemporaries in its stream.

Tyndall was also in touch with Louis Pasteur himself, who had been conducting similar investigations. A year after Tyndall's address to the Royal Society, in 1877, Pasteur and his colleague Jules-François Joubert noticed that the growth of anthrax bacilli was inhibited by mold. "Life hinders life," Pasteur wrote. "A liquid invaded by an organized ferment, or an aerobe, makes it difficult for an inferior organism to multiply." Pasteur and Joubert, though, do not seem to have investigated penicillin per se.

Twenty-five years after this rash of activity, the great wheel of rediscovery turned again, dramatically. In 1897 a young medical student in Lyon, Ernest Duchesne (1874–1912), submitted in partial fulfillment for the requirements of his doctoral degree a thesis entitled *"Contribution à l'étude de la concurrence vitale chez les micro-organismes. Antagonisme entre les moisissures et les microbes,"* or "A contribution to the study of vital competition among microorganisms. Antagonism between molds and microbes." André Maurois, surveying the prehistory of penicillin, duly notes Duchesne's conclusion: "It is to be hoped that if we pursue the study of biological rivalry between moulds and microbes, we may, perhaps, succeed in discovering still other facts which may be directly applicable to therapeutic science." Yet Maurois, having just dismissed Joseph Lister's work as unpursued, does this same of Duchesne: "In this case, too, the search was not continued."

The statement is an astounding one, though whether it results from an attempt to glorify Fleming at the expense of forgotten ancestors or mere carelessness is difficult to say. (The Panopticon does boast the world's largest collection of cited but unread works.) For had Maurois noticed, in the short span of twenty pages sandwiched between Duchesne's thesis and conclusion, the author describes several extraordinary experiments. Among them are several in which he inoculates guinea pigs with lethal amounts of bacteria along with doses of the *Penicillium glaucum* he had cultivated. Duchesne watches as the animals become sick to the point of death, then miraculously recover. Fleming himself, it should be said, never performed an experiment with live animals, and the similar one carried out by Howard Florey and Ernst Chain forty-five years later has been termed "one of the most crucial experiments in the history of medicine."

Duchesne's research was apparently ignored, but in no sense does that devalue his work. As to why his "study of biological rivalry between moulds and microbes" went unpursued, it is hard to imagine that in 1897, let alone in 1876, anyone could conceive of mass-producing penicillin. In any case, one reason is clear: Duchesne himself died tragically of tuberculosis in 1912.

How much of this history Fleming knew is anyone's guess. It can't be denied that since World War I he had been consciously searching for antibacterial agents. In their book *Les Associations microbiennes,* Papacostas and Gaté discuss literally hundreds of cases of molds vs. bacteria. Penicillin makes several appearances; Duchesne gets more than one mention (though the Englishmen none at all). It's unknown whether Fleming had read the copy that sat on the dusty shelves of the Royal Society of Medicine, of which he was a fellow, though he was aware of the general fund of work on the antibacterial properties of molds—a fund without which his great discovery would never have been made.

But let us turn to what did not happen after Fleming. The main gap in the standard tale amounts to about twelve years, from 1928, the year of the discovery, to 1941, when the first clinical trials of penicillin were made. Contrary to Maurois' account of a lonely figure struggling to get his discovery brought to the public, Fleming's view of penicillin was almost entirely scientific. He had briefly investigated its role as a local antiseptic but never injected the stuff into live animals, and he more or less abandoned the study of penicillin altogether within a year after discovering it. He played virtually no role in transforming penicillin into a therapeutic agent.

Fleming's work, in fact, lay largely forgotten until 1938, when an Oxford research team headed by Howard Florey and Ernst Chain decided to investigate antibacterial agents and came across Fleming's paper during their literature search. Procuring a sample of penicillin from the lab down the hall, Chain soon realized he had a remarkable substance on hand. It was Florey, the great organizer, who lobbied and petitioned, and who finally received a grant from the Rockefeller Foundation in order to attempt purification and production of the drug. In 1940, with 100 milligrams of penicillin on hand, Chain impetuously had a colleague inject two mice with the stuff, and it proved nontoxic. On May 25, as the battle of Dunkirk raged across the Channel, Florey and Chain did the first controlled experiments involving penicillin versus streptococci. As Duchesne had done forty-odd years earlier, they injected mice with bacteria and penicillin. And like Duchesne before them, they witnessed a miracle.

In early 1941 they tried penicillin on the first human subject—a policeman fighting a bacterial infection that had begun with a scratch on his face from a rosebush. The infection had spread so relentlessly over two months that sulfa drugs had failed to help him, his lungs and shoulders had become infected, and his left eye had been removed. By the time the Oxford team got to him, he probably had only days to live. The treatment resulted in dramatic improvement within twenty-four hours. Tragically, however, the supply of penicillin was limited—even as it was recovered from the patient's urine—and when it ran out, he suffered a relapse and died.

Nevertheless, the early clinical trials were such an extraordinary success that mass production soon ramped up, mostly in America for the war effort, and by 1942 the wonder drug was drawing attention on both sides of the Atlantic.

Publicity did them in. Relations between Fleming and the Oxford team began well enough. Getting wind of the Oxford work through the journal *The Lancet,* Fleming paid Florey and Chain an impromptu visit in 1940. When Florey announced his visit the day before, Chain was taken aback—he had not realized Fleming was still alive! Nevertheless, the meeting was cordial; the Oxonions presented Fleming with a sample of their best preparation, whose ancestor via the "lab down the hall" had been Fleming's own. Thereafter letters were exchanged and Florey kept Fleming abreast of their progress. The friction set in once the clinical trials began attracting the attention of the press. When an editorial appeared in the *British Medical Journal* highlighting the Oxford work, Fleming replied, emphasizing his priority. When the *Times* mentioned Oxford in connection with a new wonder drug, Sir Almroth Wright, Fleming's supervisor, wrote a famous letter suggesting—demanding—that credit for the discovery go to Fleming. The following day at St. Mary's Hospital, Fleming was besieged.

It couldn't have hurt that only a few weeks earlier, so as not to be left behind, Fleming had saved a man's life by treating him with penicillin provided by Florey, a recovery witnessed by the entire hospital. While Fleming indulged the press, Florey, unaware of the supreme maxim, "Without publicity there is no prosperity," spurned it. As a result, the glory accrued to Fleming, and the others were gradually forgotten. In a rare show of wisdom the Nobel committee awarded the 1945 prize for medicine to Fleming, Florey, and Chain, but by that time it was too late; the press usually forgot to mention two of them.

The saga of penicillin holds much truth. Many, in surveying this brief outline, will reply that the fact that Fleming was not the first to discover penicillin in no way detracts from his realization that his discovery was worth hanging on to. Yes, but the reverse also holds. That Fleming's discovery occurred in an age when the role of bacteria was understood, and when mass production of the drug soon became possible, in no way diminishes the research of his predecessors. It also raises the question, which perhaps has been raised before, of whether awards and honors should be given for accidental discoveries. The importance of a discovery often remains unclear for years, if not decades, and it takes no more—or less— talent to make a critical accidental discovery than a useless one. It does take more luck. In the case of penicillin, Fleming's discovery would have remained of academic interest had Florey and his team not made the clinical trials and ensured mass production of the drug. In this sense, the story should perhaps belong in the Technology Domain of the Panopticon, where it is less important who does something first than who does it best. And who has the money.

The final lesson, though, hangs in the rain-washed air. If Florey and his colleagues were largely ignored by the general public, at least they could take solace in the fact that with the limelight on Fleming, they had time to get some work done.

IV

The Domain of Mathematics: Closed for Renovation

Frequent Panopticon visitors have been asking why today's excursion did not include a stop in the Domain of Mathematics. As you see, the domain is currently closed for renovation, which will continue indefinitely because the number of its holdings is uncountable. We can give only the briefest hint of the treasures to be found there, but rest assured that mathematicians are second to none in the uncritical acceptance of their own mythology. The best proof of this must be the old chestnut about why there is no Nobel prize in mathematics.* Ask any mathematician or scientist and the likely explanation will be (unless continued debunking of the matter has laid it to rest) that Gösta Mittag-Leffler, the leading Swedish mathematician of the day and fifteen years younger than Nobel, had had an affair with Nobel's wife. Enraged, Nobel refused to endow a mathematics prize to be certain that Mittag-Leffler wouldn't win it. The trouble with this version of the tale is that Nobel was a "retiring, considerate person who detested all forms of publicity." And a confirmed bachelor. On the other hand, a Finnish mathematician I used to know once claimed to have visited Mittag-Leffler's house in Stockholm, and in the back garden was a bust of Nobel's sister . . . Funnily enough, the Nobel prize historian Elisabeth Crawford actually gives some credence to the whole story by conjecturing that Mittag-Leffler, in revenge for the lack of a mathematics prize, actually spread the rumor himself; Nobel had lost out in their rivalry for a woman and so refused to create the prize. Maybe. Mittag-Leffler was a malicious gossipmonger, and the two men evidently hated each other.

Then there is the famous story of the child Carl Friedrich Gauss, by common consent the greatest mathematician of all time. You know the one I mean: His teacher, annoyed at the unruly class, punished his charges

---

* The Norwegian government has recently endowed a Nobel prize for math, to be called the Abel Prize, inaugurated in 2003.

by having them add up all the numbers from one to one hundred. Gauss almost instantly derived a summation formula in his head, chalked the answer on his slate, and placed it facedown on the desk with the triumphant declaration, "There it lies." Gradually the other students finished and piled their slates atop Gauss'. One by one the teacher removed them. All the answers were incorrect except Gauss'. The story has no basis in fact.

For the celebrated legends surrounding the life and death of firebrand Evariste Galois, a pioneer of group theory who was killed in a duel at the age of twenty, I refer you to my own article in *Science à la Mode* or on the Web. No prostitutes or agents provocateurs were involved, alas, and Galois didn't create group theory the night before he was shot. A reliable source tells me that a similar article exists showing that Niels Heinrich Abel, famous for his proof that it is impossible to solve a general fifth-degree equation by algebraic means, did not die in poverty, as universally believed; however, I have not been able to locate it. Newton's binomial theorem was known centuries before Newton. An important untold story in mathematics is that certain aspects of chaos theory, thought to have originated in the 1970s, were actually worked out in 1945 by Mary Cartwright and John E. Littlewood—without computers.

One could go on infinitely. However, life is too short. We are thus grateful for the restoration work on the Panopticon and are able to take our leave with only a few final words. Apart from those listed in the preface, the main lesson to emerge from this meander is the most obvious one: Science is a collective endeavor. Even the greatest scientists have borrowed, begged, and stolen the ideas surrounding them. As John Wheeler used to remark, the definition of a scientist is "an unscrupulous opportunist," which is often true in more ways than he probably meant. For that reason, to repeat myself somewhat, it becomes increasingly apparent with every tale that the journalistic biography, so much the fashion for several decades now, is not the way to go. In focusing on one protagonist, the contributions of others are invariably diminished or ignored, and the result is a reinforcement of the long-obsolete notion that science proceeds by the genius of isolated individuals percolating their inspiration into the vacuum. It is a totally warped view of the true state of affairs. The protagonist should be the ideas.

Additionally disturbing, to take up the theme with which this journey began, is that the narration becomes its own imperative. As the adrenaline gets revved up, the story takes off. How rarely we encounter a biographer interrupting the flow of events with the concession "Something is missing; the facts can't be merged into a consistent account." Rather, as the plot line takes on a life of its own, gaps get papered over to produce a seamless thriller, which may be as misleading as it is entertaining. And as we've all been weaned on Hollywood, the story tends to becomes an all-or-nothing affair. Either Bell invented the telephone or he didn't. Either Farnsworth invented the television or he didn't. Why one can't merely say this person did this and that person did that escapes me. Of all the things to say, the second most important is, when possible, how one person influenced the next, how the competitors interacted. Most important is to say, when necessary, "I don't know."

Typically, scientists view the history of science as a quaint, amusing pastime, an indulgence only over-the-hill scholars can afford, yet it seems to me that the reduced history we are all exposed to does have some consequences for the progress of science and education. That is because the scientific community at large has partaken of the image it has created of itself, and it is self-serving. Each academic department and each research division believes it knows what is important, who is doing the good work and who isn't. Given the number of mistakes that have occurred along these lines, one might have expected a little more humility to emerge, but every field continues to lurch from fashion to fashion, convinced at each cycle that it has taken the one and true path. At the Contemporary Panopticon of Present and Past Concepts a perpetual war is waged between the idea that one should do science for the good of the greater community and the conviction that science, like art, should be pursued for personal enrichment. Certainly the latter stand has forever been on the defensive; the establishment has never permitted much useless self-indulgence. Nowadays the pressure to do something "practical" is so intense that many students, convinced they can make no worthwhile contribution to their field, drop out, and those that remain do so for largely mercenary reasons. The atmosphere has been unquestionably cheapened, which is too bad. And while the achievement of fame and fortune in science is not a reasonable goal for most researchers, at least the stories of unsung pioneers such as Niepce de Saint-Victor or Kaufmann or Loschmidt can give some solace. Even if you haven't gotten the credit, you've done the calculation, and not only has the calculation itself bestowed some understanding, but it occasionally makes a difference.

REFERENCES AND NOTES

I. The Domain of Physics and Astronomy

Introduction

The classic criticism of Viviani's account of Galileo's experiment is Lane Cooper, *Aristotle, Galileo and the Tower of Pisa* (Ithaca, Cornell University Press: 1935).

See also Michael Segre, "The Never-ending Galileo story," in Peter Machamer, ed., *Cambridge Companion to Galileo* (Cambridge, Cambridge University Press: 1998); Michael Segre, "Galileo, Viviani and the Tower of Pisa," *Stud. Hist. Phil. Sci.* **20,** 435 (1989); Michael Segre, *In the Wake of Galileo* (New Brunswick, Rutgers University Press: 1991).

2 Stevin's experiment is discussed by Lane; Viviani's description of the Leaning Tower experiment is as cited by Segre, "Galileo," p. 437. For Viviani's notes attempting to link Michaelangelo and Galileo, see Segre, *In the Wake,* pp. 116–22.

3 For more on superconductivity, see Jacobus de Nobel, "The Discovery of Superconductivity," *Physics Today,* Sept. 1996, p. 40; and Rudolf de Bruyn Ouboter, "Heike Kamerlingh Onnes' Discovery of Superconductivity," *Scientific American,* March 1997, p. 98.

W. Ehrenberg and R. E. Siday, *Proc. Phys. Soc.* (Great Britain), **B62,** p. 8 (1949).

1. The Mafia Invents the Barometer

The account of Torricelli is largely based on the excellent study by W. E. Knowles Middleton, *The History of the Barometer* (Baltimore: Johns Hopkins University Press, 1964), chapters 1–3. The translations of Torricelli's letters are as given by Middleton.

5 "Other units in common use . . ." is from Hans C. Ohanian, *Physics* (New York: W. W. Norton, 1985), p. 451.

"Another instrument in common use . . ." is from Raymond A. Serway, *Physics for Scientists and Engineers* (Philadelphia: W. B. Sanders, 1992), p. 399.

Berte Bolle, *Barometers* (Herts, England: Argus Books, Ltd., 1982), p. 12.

Sheldon Glashow, *From Alchemy to Quarks* (Cambridge, Mass.: Harvard University Press, forthcoming). (I pointed out the confusion to Glashow in a prepublication edition some years ago. Perhaps the mistake has been corrected.)

6 Isaac Asimov, *Chronology of Science and Discovery* (New York: HarperCollins, 1994), p. 164.

"In 1643, Torricelli proposed his experiment, which was carried out by his colleague Viviani" was found at http://www.gap.dcs.st-and.ac-uk/~history/Mathematics/Torricilli .html.

2. The Riddle of the Sphinx: Thomas Young's Experiment

A good biography of Young is Alexander Wood with Frank Oldham, *Thomas Young, Natural Philosopher* (Cambridge: Cambridge University Press, 1954).

Young's famous lectures are *A Course of Lectures on Natural Philosophy and the Mechanical Arts,* 2nd ed. (London: Taylor and Walton, 1835).

See also G. N. Cantor, *Optics After Newton* (Manchester: Manchester University Press, 1983) and G. N. Cantor, "Was Thomas Young a Wave Theorist?" *Am. J. Phys.* **52,** 305–8.

13 Details of Young's childhood can be found in Wood with Oldham, chap. 1.

15 Young's remarks on Newton's theory are from *A Course of Lectures,* Lecture XXXIX, p. 362.

Newton's reply to Hooke is *Phil. Trans. Roy. Soc.* **7,** 5084–103 (1672), p. 5086.

17 The 1801 Bakerian Lecture is *Phil. Trans. Roy. Soc.* **92,** 12–49 (1802).

17–18 The 1800 paper on sound and light can be found in *Phil. Trans. Roy. Soc.* **90,** pp. 106–55 (1800).

The 1802 lecture in which Young clearly announces the principle of interference is *Phil. Trans. Roy. Soc.* **92,** pp. 387–97 (1802).

The 1803 Bakerian Lecture is *Phil. Trans. Roy. Soc.* **94,** pp. 1–16 (1804).

18–19 The quotations from Young on the famous "experiment" are as found in *A Course of Lectures,* Lecture I, p. 1; Lecture XXIII, pp. 219–20; Lecture XXXIX, pp. 364–65. Young's remarks on what he actually did are from the 1803 Bakerian Lecture, p. 2.

John Worrall's broadside on Young is "Thomas Young and the 'Refutation' of Newtonian Optics" in Colin Howson, ed., *Method and Appraisal in the Physical Sciences* (Cambridge: Cambridge University Press, 1976).

21 More on Young and the Rosetta Stone can be found in Wood and Oldham, chaps. 9–10.

3. Joseph Henry and the (Near) Discovery of (Nearly) Everything

Much of the material for this chapter is based on *The Papers of Joseph Henry* (Washington, D.C.: Smithsonian Institution Press, 1972–85), volumes 1–4, as well as *The Scientific Writings of Joseph Henry* (Washington, D.C.: Smithsonian Institution Press, 1886).

Also helpful have been the two standard biographies of Henry by Thomas Coulson, *Joseph Henry* (Princeton: Princeton University Press, 1950), and Albert Moyer, *Joseph Henry: The Rise of an American Scientist* (Washington, D.C.: Smithsonian Institution Press, 1997). The latter is of course more up-to-date with access to more documents, but is written in a far more academic style. The essentials are the same.

Some material relating to Henry's inventions and a few of his letters are now available at the Joseph Henry Project Web site maintained by the Smithsonian (www.si.edu/archives/ind/jhp).

25 The inscription in Gregory is from *Papers of Joseph Henry,* Volume I, p. xxii.

26 Description of Niagara is from ibid., p. 149.

26–27 Inaugural address is from ibid., p. 178.

28 For Romagnosi, see John J. Fahie, *A History of Electric Telegraphy to the Year 1837* (New York: Arno Press, 1974 [1884]), pp. 257 ff.

Letter to Beck, *Papers of Joseph Henry,* p. 196.

29 Henry on Moll, *Scientific Writings,* vol. I, p. 49.

30 Henry on Ritchie, *Papers of Joseph Henry,* vol. II, p. 162.

31 Henry on Faraday, *Scientific Writings,* vol. I, p. 75. On self-induction, p. 79.

32 Henry in *Encyclopedia Americana,* as quoted by Moyer, p. 175. For electric eels, see *Scientific Writings,* vol. I, p. 161; for sparks from attic to cellar and the "diffusion of motion," see p. 203.

4. Neptune: The Greatest Triumph in the History of Astronomy, or the Greatest Fluke?

A detailed popular account of the Neptune story can be found in Tom Standage, *The Neptune File* (New York: Berkeley Books, 2000), which has provided much useful historical information for this retelling.

A splendid scientific account is Benjamin Apthorp Gould, *Report to the Smithsonian Institution on the History of the Discovery of Neptune,* (Washington, D.C.: Smithsonian Institution Press, 1850).

A detailed historical account of the scientific controversy surrounding Peirce's claims can be found in John Hubbell and Robert Smith, "Neptune in America: Negotiating a Discovery," *Journal for the History of Astronomy* **23,** pp. 261–91 (1992). The authors seem determined not to arrive at any conclusions regarding the truth of Peirce's arguments.

Peirce's own remarks are given at length in the *Proceedings of the American Academy of Arts and Sciences* **1,** 1846–48. See sessions for March 16, 1847; May 4, 1847; December 7, 1847; April 4, 1848.

35 William Kaufmann, *Universe,* 4th ed. (New York: W. H. Freeman, 1994), pp. 272–73.

36 Eric Chaisson and Steve McMillan, *Astronomy Today* (Englewood Cliffs: Prentice Hall, 1993), p. 308.

"With respect to this planet . . .": Standage, p. iv.

38 "I have already begun to mount the ladder . . .": ibid., p. 95.

39 "Analysis transports us to the regions of the unknown . . .": ibid., p. 113.

40 "It must, indeed, be confessed . . .": Gould, p. 19. Challis is as quoted by Gould.

41 "That star is not on the map!": Standage, p. 120.

"Le Verrier is called upon today to share the glory . . .": ibid., p. 134.

42 "it was thus rendered a moral certainty . . .": Gould, p. 43.

"From these data . . .": Peirce, *Proc. Amer. Acad.* **1,** p. 65 (March 16, 1847, session). Peirce's March criticism concerning the incorrect estimates of Neptune's mass and distance was not his only one. At the May 4, 1847, session, Peirce also pointed out that two solutions to the problem other than the one found by Adams and Le Verrier existed, each of which put Neptune about 120° away in the sky from where it was actually found. "If the above geometers had fallen upon either of these solutions instead of the one which was obtained," Peirce concluded, "Neptune would not have been discovered in consequence of geometrical prediction." This point is not so crucial as the one concerning Neptune's distance and mass; Adams and Le Verrier could have legitimately predicted that the New Planet had to be in one of three positions in the sky.

43 "Candor and moral courage . . .": Gould, p. 44.

Edward Everett's and Joseph Henry's remarks are in Hubbell and Smith, p. 275.

44 Le Verrier's arguments are as summarized by Gould, p. 53.

Herschel's remarks can be found in his *Outlines of Astronomy* (London: Longman, 1849), p. 516.

5. Invisible Light: The Discovery of Radioactivity

Insofar as possible, this chapter has been based on the original memoirs of Niepce de Saint-Victor, Eugène Chevreul, Henri Becquerel, and Gustave Le Bon that appeared in the *Comptes Rendus* of the French Academy of Science. They are:

Niepce de Saint-Victor, Première Mémoire, "Sur une nouvelle action de la lumière," *C. R. Acad. Sci.* **45**, pp. 811–15 (1857).

Niepce de Saint-Victor, Deuxième Mémoire, *C. R. Acad. Sci.* **46**, pp. 448–52 (1858).

Niepce de Saint-Victor, Troisième Mémoire, *C. R. Acad. Sci.* **47**, pp. 866–69 (1858).

Niepce de Saint-Victor, Quatrième Mémoire, *C. R. Acad. Sci.* **47**, pp. 1002–6 (1858).

Niepce de Saint-Victor, Note sur l'activitè, *C. R. Acad. Sci.* **48**, p. 741 (1859).

Niepce de Saint-Victor, De l'action, *C. R. Acad. Sci.* **49**, pp. 815–17 (1859).

Niepce de Saint-Victor, Cinquième Mémoire, *C. R. Acad. Sci.* **53**, pp. 33–39 (1861).

Niepce de Saint-Victor, Sixième Mémoire, *C. R. Acad. Sci.* **65**, pp. 505–11 (1867).

M. E. Chevreul, *C. R. Acad. Sci.* **47**, pp. 1006–11 (1858).

H. Becquerel, *C. R. Acad. Sci.* **122**, pp. 420–21 (1896).

H. Becquerel, *C. R. Acad. Sci.* **122**, pp. 501–3 (1896).

H. Becquerel, *C. R. Acad. Sci.* **122**, pp. 559–64 (1896).

H. Becquerel, *C. R. Acad. Sci.* **122**, pp. 689–94 (1896).

H. Becquerel, *C. R. Acad. Sci.* **120**, pp. 762–68 (1896).

H. Becquerel, *C. R. Acad. Sci.* **122**, pp. 1086–88 (1896).

H. Becquerel, *C. R. Acad. Sci.* **128**, pp. 771–74 (1899).

G. Le Bon, *C. R. Acad. Sci.* **122**, pp. 188–90 (1896).

G. Le Bon, *C. R. Acad. Sci.* **122**, pp. 233–35 (1896).

G. Le Bon, *C. R. Acad. Sci.* **122**, pp. 386–90 (1896).

G. Le Bon, *C. R. Acad. Sci.* **122**, pp. 462–63 (1896).

G. Le Bon, *C. R. Acad. Sci.* **122**, pp. 522–24 (1896).

G. Le Bon, *C. R. Acad. Sci.* **124**, pp. 755–58 (1897).

G. Le Bon, *C. R. Acad. Sci.* **124**, pp. 892–95 (1897).

G. Le Bon, *C. R. Acad. Sci.* **124**, pp. 1148–51 (1897).

Also helpful have been:

P. Fournier and J. Fournier, *New J. Chem.* **14**, pp. 785–90 (1990).

M. Meyer and E. Gonthier, "Y a-t-il encore polémique autour de la découvert des phénomènes dit radioactifs?" *Science Tribune* (July 1997); www.tribunes.com/tribune/art97/meyer.htm.

G. Le Bon, *L 'Evolution de la matière* (Paris: Flammarion, 1908), chaps. 2 and 14.

47 "Does a body . . . ?": Niepce de Saint-Victor, Première Mémoire, p. 811.

48 "I have observed . . .": Niepce de Saint-Victor, Deuxième Mémoire, p. 452.

"The facts set forth . . .": Chevreul, p. 1009.

49 "The persistent activity . . .": Niepce de Saint-Victor, Cinquième Mémoire, p. 34.

"the action of light is perhaps very favorable on certain wines . . .": Niepce de Saint-Victor, "De l'action," p. 817.

50 For de Heen's pamphlet and Le Bon's side of the story, see Le Bon, *L 'Evolution*, chap. 14. For his remark implying that Becquerel knew of Niepce's discovery, see chap. 2, p. 24.

For Becquerel's side, see J. Becquerel, *Les inventeurs célèbres, sciences physiques et applications* (Paris: Mazenod, 1962), pp. 298–99.

6. Light, Ether, Corpuscles, and Charge: The Electron

A recent book that is sure to become required reading on the electron, and that has been extremely useful for this chapter, is *Histories of the Electron: The Birth of Microphysics*, edited by Jed Buchwald and Andrew Warwick (Cambridge, Mass.: MIT Press, 2001). See in particular the chapters by George Smith, Isobel Falconer, Theodore Arabatzis, Helge Kragh, and Peter Atchinstein.

Also useful have been:

Jost Lemmerich, "The History of the Discovery of the Electron," lecture delivered at the Lepton Photon Symposium 1987 (courtesy of the author).

"The Discovery of the Electron," *Highlights from the DESY Research Center*, 1998, pp. 35–39.

Abraham Pais, *Inward Bound* (New York: Oxford University Press, 1986), chap. 4.

Sir Edmund Whittaker, *A History of the Theories of Aether and Electricity* (New York: American Institute of Physics, 1987).

55 Faraday's remarks are from *Phil. Trans. Roy. Soc.* **124**, pp. 116, 121 (1834).

G. Johnstone Stoney, *Phil. Mag.* **11**, pp. 381–90 (1881); *Phil. Mag.* **38**, pp. 418–20 (1894). Perhaps even more remarkably than his introduction of the "electron" is that in the first paper Stoney introduced a set of "natural units," which are identical to the Planckian units in universal use among physicists today, except that Stoney's were based on the speed of light, the gravitational constant, and the charge of the electron, whereas Planck substituted Planck's constant for the last.

56 For Hughes, see Whittaker, p. 323, or "Prof. D. E. Hughes' Researches in Wireless Telegraphy," *The Electrician*, May 5, 1899, pp. 40–41. I have placed this article on my Web site (wwwrel.ph.utexas.edu/~tonyr). Hughes' records of his experiments are in the British Museum.

57 Wilhelm Hallwachs, *Annalen der Physik und Chemie* **33**, pp. 301–12 (1888).

For the Italian Righi's photoelectric experiments, see S. Galdabini, G. Giuliani, and N. Robotti, "Photoelectricity Within Classical Physics" (Internet preprint, http://matsci.unipv.it/percorsi/giuliani_pubb.htm).

58 Hertz is as quoted in Albrecht Fölsing, *Albert Einstein* (New York: Penguin Books, 1997), p. 159.

Larmor's electron is as described by Arabatzis in *Histories of the Electron*, p. 183.

59 Pieter Zeeman, *Phil. Mag.* **43**, pp. 226–39 (1897); **44**, pp. 55–60 (1897); **44**, pp. 255–59 (1897). I presume a misprint in the third paper, p. 256, in which Zeeman gives $e/m = 1.6 \times 10^{-10}$ electromagnetic c.g.s. units and says that "the order of magnitude of e/m is entirely the same as the one formerly given," i.e. 10^7.

61 Jean Perrin, *Comptes Rendus* **121**, pp. 1130–34 (1895).

Weichert's remarks are from his lecture before the Physikalisch-ökonomischen Gesellschaft, Königsberg, January 7, 1897 (courtesy Jost Lemmerich). See also Pais, p. 82.

W. Kaufmann, *Ann. der Phys. und Chem.* **61**, pp. 544–52 (1897).

W. Kaufmann and E. Aschkinass, *Ann. der Phys. und Chem.* **62,** pp. 588–95 (1897).
W. Kaufmann, *Ann. der Phys. und Chem.* **62,** pp. 596–98 (1897).

61 P. Lenard, *Ann. der Phys. und Chem.* **64,** pp. 279–89 (1898).
J. J. Thomson, *Phil. Mag.* **44,** pp. 293–316 (1897); **46,** pp. 528–45 (1898); **48,** pp. 547–67 (1899).

62 W. Kaufmann, "The Development of the Electron Idea," *The Electrician,* Nov. 8, 1901, pp. 95–97.

7. Einstein's Miraculous Year (and a Few Others)

The following have provided much general background for this chapter and the next:
Abraham Pais, *Subtle Is the Lord* (New York: Oxford University Press, 1982).
Albrecht Fölsing, *Albert Einstein* (New York: Penguin Books, 1997).
Jagdish Mehra, "The Historical Origins of the Special Theory of Relativity," in *The Golden Age of Theoretical Physics,* vol. 1 (Singapore: World Scientific, 2001).
All excerpts from Einstein's letters are from the translations in volume 1 of *The Collected Correspondence of Albert Einstein* (Princeton: Princeton University Press, 1987), henceforth PUPC.

68 Louis Bachelier, "Theorie de speculation," *Annales Scientifique de L 'Ecole Normale* **16,** p. 21 (1900). Translated by A. James Boness in *The Random Character of Stock Market Prices,* ed. Paul H. Cootner (Cambridge, Mass.: MIT Press, 1964).

69 Hertz as quoted in Fölsing, p. 159.

72 For Einstein's youthful essay, see "Einstein's First Paper" in Mehra.
Einstein to Mileva, August 1899, PUPC, p. 131.
Einstein to Mileva, December 1901, PUPC, p. 186.
For Poincaré on absolute space, see his *Science and Hypothesis* (New York: Dover, 1952), p. 90.

73 Poincaré's technical papers are H. Poincaré, *Comptes Rendus de l'Académie des Sciences* **140,** p. 1504 (1905); *Rendiconti del Circulo Matematico di Palermo* **21,** p. 129 (1906).

74 My unpublished paper on aberration is "Reference Frames for Stellar Aberration," by Milton Rothman, Tony Rothman, and Peter Aninnos. It is available on the Web as a National Center for Supercomputing Applications (NCSA) preprint (http://archive.ncsa.uiuc.edu/Pubs/TechReports/TechReportsSeries.html).

75 "Ether and the Theory of Relativity," May 5, 1920, at the University of Leyden, published as *Sidelights on Relativity* (New York: Dutton & Co., [1922?]).
For who first derived E = mc² in generality, see Mehra, p. 214.

76 Rutherford and Soddy are as quoted by Richard Rhodes, *The Making of the Atomic Bomb* (New York: Simon and Schuster, 1986), pp. 43–44.
"What a strange thing . . . !": Einstein to Julia Niggli, August 1899, PUPC, p. 129.

8. What Did the Eclipse Expedition Really Show?
And Other Tales of General Relativity

The account of the eclipse expeditions is largely based on John Earman and Clark Glymour, "Relativity and Eclipses: The British Eclipse Expeditions of 1919 and Their Predecessors," *Historical Studies in the Physical Sciences* **11,** 1, pp. 49–85 (1980).

PUPC refers to *The Collected Papers of Albert Einstein,* vol. 8 (Princeton: Princeton University Press, 1998).

77 The newspaper headlines are as in Abraham Pais, *Subtle Is the Lord* (New York: Oxford University Press, 1982) and Albrecht Fölsing, *Albert Einstein* (New York: Penguin Books, 1997).

79 Soldner's paper has been reprinted with an introduction by Stanley Jaki in *Foundations in Physics* **8,** p. 927 (1978).

81 The paper of Johannes Droste is in *Proceedings of the Royal Dutch Society (Kon. Akad. van Wetenschappen)* **19,** p. 197 (1917).
    Einstein on Le Verrier is from Pais, p. 253.

82 The unpublished version of general relativity by Einstein and Besso is in PUPC, German text 3, pp. 361–473 (with English editorial notes).

83 Ilse to Nicolai is PUPC, English text 8, p. 564.

86 Eddington's main popular account of the expeditions can be found in his *Space, Time and Gravitation* (Cambridge, Mass.: Cambridge University Press, 1921).

87 Einstein's words to Besso on his theory are from Pais, p. 303.
    Einstein to Sommerfeld is PUPC, 8 p. 153.
    Einstein to Ehrenfest is from Albrecht Fölsing, *Albert Einstein* (New York: Penguin Books, 1997), p. 439.

9. Two Quantum Tales: Bohr and Hydrogen, Dirac and the Positron

The standard histories on the Nicholson-Bohr connection are:

Russell McCormmach, "The Atomic Theory of John William Nicholson," *Archive for History of Exact Sciences* **3,** pp. 160–84 (1966).

John Heilbron and Thomas Kuhn, "The Genesis of the Bohr Atom," *Historical Studies in the Physical Sciences,* **1,** pp. 211–91 (1969).

Leon Rosenfeld, introduction to Niels Bohr, *On the Constitution of Atoms and Molecules,* papers of 1913 reprinted from the *Philosophical Magaine* (New York: W. A. Benjamin, 1963).

For a contra-Nicholson history, see Tetu Hirosige and Sigeo Nisio, "Formation of Bohr's Theory of Atomic Constitution," *Jap. Studies Hist. Sci.* **3,** pp. 6–28 (1964). I believe their conclusions to be incorrect.

Dirac's lecture was reprinted as P. A. M. Dirac, *The Development of Quantum Theory* (New York: Gordon and Breach, 1971).

88 Raymond Serway, Clement Moses, and Curt Moyer, *Modern Physics,* 2nd. ed. (New York: Harcourt, 1997), p. 131.

92 Nicholson's papers are *Mon. Not. Roy. Astr. Soc.* **72,** p. 50 (1911); **72,** pp. 139–50 (1911); **72,** pp. 677–92 (1912); **72,** pp. 729–39 (1912). The quotations are from the third paper, pp. 677, 679.

94 Jeans is as quoted by McCormmach, p. 184.
    Rosenfeld is as quoted in Rosenfeld, pp. xiii–xiv.
    For Pais' view, see Abraham Pais, *Niels Bohr's Times* (New York: Oxford University Press, 1991), pp. 145–47.
    "All Bohr had to do . . .": Jagdish Merha and Helmut Rechenberg, *The Historical Development of Quantum Theory,* vol. 1, part 1 (New York: Springer-Verlag, 1982), p. 189.

97 Dirac's theory of electrons and protons is in *Proc. Roy. Soc. Lon.,* Series A **126,** pp. 360–65 (1930). The extract from his lecture is to be found in *Development of Quantum Theory,* pp. 51–55.

10. A Third Quantum Tale: Southpaw Electrons and Discounted Luncheons

Fairly complete accounts of the V–A episode can be found in:

Jagdish Mehra, *To the Beat of a Different Drum: The Life and Science of Richard Feynman* (Oxford: Clarendon Press, 1994), chap. 21.

N. Mukunda, "E. C. G. Sudarshan and the Development of Weak Interaction Theory," *Current Science* **63**, p. 59 (July 25, 1992).

Robert Marshak, "The Pain and Joy of a Major Scientific Discovery," *Current Science* **63**, pp. 60–64 (July 25, 1992).

E. C. G. Sudarshan, "Mid-century Adventures in Particle Physics," preprint, Center for Particle Theory, University of Texas, Austin, July 1985.

An abridged account, almost entirely from Feynman's point of view, can be found in James Gleick, *Genius* (New York: Pantheon, 1992), pp. 337–38.

98 The quote of Feynman is from Mehra, p. 453.

100 For more on the weak force and parity conservation, see Tony Rothman and George Sudarshan, *Doubt and Certainty* (Cambridge, Mass.: Perseus, 1998), chap. 6.

Pauli is as quoted by Burton Feldman, *The Nobel Prize* (New York: Arcade Publishing, 2000), p. 172.

101 Feynman's version is as told to Mehra, pp. 464–67.

103 For Feynman's apology to Marshak, see Mehra, pp. 477–78, and Marshak, p. 63.

Bethe is as quoted by Marshak himself, p. 63.

Gell-Mann's "It seems an unreasonable conclusion . . ." is from Mehra, p. 435.

II. The Domain of Technology

Introduction

The main biography of Gutenberg available in English is Albert Kapr, *Johann Gutenberg* (Brookfield, Vt.: Ashgate Press, 1996). This is a complete, scholarly account, which I think demonstrates how little is really known about its protagonist.

A treasure trove of information about steam engines can be found on the Web at the Steam Engine Library, maintained by the University of Rochester (www.history.rochester.edu/steam).

A succinct and lucid book disputing much of the information in the Steam Engine Library is H. Philip Spratt, *The Birth of the Steamboat* (London: Charles Griffin, 1958).

105 For the Coster legend, see Kapr, chap. 4.

106 For the recent digital work by Needham and Agüera y Arcas, see the *Princeton Weekly Bulletin* for Feb. 12, 2001, available at www.princeton.edu/pr/pwb/01/0212/.

107 Full accounts of Savary, Fitch, and Rumsey can be found at the Steam Engine Library, including Savary's original description of the miner's friend and histories of the Fitch-Rumsey dispute. See also Spratt for all of this, especially pp. 18–19.

Some accounts of Rumsey indicate his first boat was of the barge-pole variety. I have not been able to verify this, and Spratt, who seems careful, mentions only the jet engine.

108 John MacGregor, "On the Paddle Wheel and Screw Propeller," *The Artizan* **16**, pp. 108–11 (1858). See also Spratt.

109   Daniel H. Thomas, "Pre Whitney Cotton Gins in French Louisiana," *The Journal of Southern History,* **31,** pp. 135–48 (1965).
Seale Ballenger, *Hell's Belles* (Berkeley: Conari Press, 1997), p. 189.
David R. Starbuck, "Re-Inventing Eli Whitney," *Archaeology,* Sept.–Oct. 1997, p. 100.

11. What Hath God Wrought? Shadows of Forgotten Ancestors, Samuel Morse, and the Telegraph

The standard history of the telegraph is John J. Fahie, *A History of Electric Telegraphy to the Year 1837* (New York: Arno Press, 1974 [1884]). See also T. K. Derry and Trevor I. Williams, *A Short History of Technology* (New York: Dover, 1993 [1960]).

The two biographies of Henry are, again, Thomas Coulson, *Joseph Henry* (Princeton: Princeton University Press, 1950) and Albert Moyer, *Joseph Henry: The Rise of an American Scientist* (Washington, D.C.: Smithsonian Institution Press, 1997). Coulson leans much more toward Henry in the telegraph dispute than Moyer.

A helpful biography of Morse has been Paul J. Staiti, *Samuel F. B. Morse* (Cambridge, Mass.: Cambridge University Press, 1989). There is no mention of the telegraph controversy.

The standard collection of Morse's letters is Samuel F. B. Morse, *Letters and Journals, Edited and Supplemented by His Son, Edward Lind Morse* (Boston: Houghton Mifflin, 1914). The operative word is "edited." This is not so much a collection of Morse's letters and diaries, but a biography written by his son, based on Morse's letters. Extraordinarily biased. To be treated with great care.

Many Morse documents are now available at the Library of Congress Web site, www. loc.gov.

113–14   For more details on C. M.'s telegraph and the other early telegraphs, see Fahie.
118   "I have had notice of another application for a patent by a person named *Morse*": Fahie, p. 431.
Morse on American art: *Letters and Journals,* vol. I, p. 47.
119   For more on the establishment of the National Academy of Design, see Thomas Cummings, *Historic Annals of the National Academy of Design* (New York: Kennedy Galleries, 1969).
"America is the stronghold . . .": ibid., p. 429.
120   For more on the political situation in New York in the 1830s, see Paul Gilje, *The Road to Mobocracy* (Chapel Hill: University of North Carolina Press, 1987), chap. 5. See also Leo Hershkowitz, "The Native American Democratic Society in New York City," *New York Historical Society Quarterly* **XLVI,** p. 41 (1962).
"It is a religion of the imagination . . .": *Letters and Journals,* vol. I, p. 399.
Samuel F. B. Morse, *Foreign Conspiracy Against the Liberties of the United States; The Numbers of Brutus* (New York: Leavitt, Lord & Co., 1835), pp. 21–22, 95, 127.
"If the presence of electricity . . .": *Letters and Journals,* vol. II, p. 6.
121   "Morse's machine was complete in all its parts . . .": *Letters and Journals,* vol. II, p. 54.
"an unassuming and prepossessing gentleman . . .": Coulson, p. 128.
"if the length of wire between stations is great . . .": *Letters and Journals,* vol. II, p. 141.
122   "it is not visionary to suppose . . .": ibid., p. 85.
123   "feelings of deep regret and mortification . . .": Moyer, p. 244.
"This is a very good letter . . .": Coulson, p. 218.

123 "I will have nothing to do with Mr. Vail!": Morse, *Defence Against the Injurious Deductions Drawn from the Deposition of Prof. Joseph Henry,* 1855(?), reprinted 1857(?), p. 71.

124 Morse "had made no discoveries in science . . .": report to the Smithsonian, 1857, as quoted by Fahie, p. 498.

"I claim to be the first . . .": ibid., p. 503.

"Our country is more indebted . . .": *Scientific American,* November 15, 1851, p. 67. See also letter from Tal. P. Shaffner to S. Morse, reprinted in *Defence,* p. 3.

It becomes "not less a duty to the cause of Historical truth . . .": *Defence,,* pp. 8 ff.

126 "God has chosen me as the instrument": *Letters and Journals,* vol. II, p. 267.

12. Fiat Lux: *Edison, the Incandescent Bulb, and a Few Other Matters*

For this chapter I am indebted to Neil Baldwin, *Edison: Inventing the Century* (New York: Hyperion, 1995). See also Matthew Josephson, *Edison* (New York: McGraw Hill, 1959).

An in-depth study of the development of the incandescent bulb at Menlo Park, though with scant reference to anyone else, is Robert Friedel and Paul Israel, *Edison's Electric Light: Biography of an Invention* (New Brunswick, N.J.: Rutgers University Press, 1986).

Some older books have also proven useful:

John Howell and Henry Schroeder, *History of the Incandescent Lamp* (Schenectady: Maqua, 1927).

Arthur A. Bright, *The Electric Lamp Industry* (New York: Macmillan, 1949).

128 "For two years . . .": Baldwin, p. 202.

129 "The simple fact . . .": Friedel and Israel, p. xii.

"Toying one night with . . .": *Edisonia, a Brief History of the Early Edison Electric Lighting System* (New York: Comm. of St. Louis Exposition of the Assoc. of Edison Illuminating Companies, 1904). Probably originally from the *New York Herald.*

"The Incandescent Electrical Light," *Canadian Electrical News,* **10,** 2 (1900) (pages not numbered; most of issue).

132–33 For Lodygin, see Howell and Schroeder. I have also consulted the Lodygin Family Papers at Columbia's University's Rare Book and Manuscript Library, housed at Butler Library. The information about Lodygin's company is from a manuscript, "Aleksandr Nikolaevich Lodygin," which is unsigned but apparently written by his daughter. The file also contains both English and Russian versions of Albert Parry's series "The Legendary Lodygin," which appeared in *Novoe Russkoye Slovo,* October 21–November 1, 1972. The claim about Khotinksy is found in the October 26 installment.

133–34 The passages from Swan, the London paper, and *La Lumière electrique* are as in Baldwin, pp. 124–35.

136 Charles Cros, *Comptes Rendus* **35,** pp. 1082–83 (1877).

13. "Magna Est Veritas et Praevalet": *The Telephone*

The Bell biographies consulted were:

Robert Bruce, *Alexander Graham Bell and the Conquest of Solitude* (Boston: Little, Brown, 1973).

Edwin Grosvenor and Morgan Wesson, *Alexander Graham Bell* (New York: Abrams, 1997), p. 44.

James MacKay, *Alexander Graham Bell: A Life* (New York: John Wiley and Sons, 1997).

For an almost orthogonal view of events, see Lewis Coe, *The Telephone and Its Several Inventors* (Jefferson, N.C.: McFarland, 1995).

An older, somewhat disorganized, but highly detailed book along the same lines is William Aitkin, *Who Invented the Telephone?* (London: Blackie and Son, 1939).

A detailed article on the Gray-Bell controversy is Lloyd Taylor, "The Untold Story of the Telephone," *American Physics Teacher* (later *American Journal of Physics*) **5**, pp. 243–51 (1937). Coe reprints this article in his book, unfortunately without the crucial references and diagrams. It is also found in Shiers (below).

The classic study of Reis' work is Silvanus Thompson, *Philipp Reis: Inventor of the Telephone* (London: E. & F. N. Spon, 1883; reprint, New York: Arno Press, 1983).

Many of the original, early articles on the telephone are collected in George Shiers, ed., *The Telephone: An Historical Anthology* (New York: Arno Press, 1977).

The court cases themselves make for exciting reading. The testimony in the major one, known as the *Dowd* case, can be found in *Telephone Suits, Circuit Court of the United States, Bell Telephone Company et al. v. Peter A. Dowd* (Boston: Alfred Mudge & Son, 1880).

Bell's testimony in this case and later ones has been collected in one volume, usually referred to as the *Deposition of Alexander Graham Bell*, published as *The Bell Telephone* (Boston: The American Bell Telephone Company, 1908).

138 Bruce, p. 177.
139 Grosvenor and Wesson, p. 44.
      Mackay, p. 89.
      Thomas A. Watson, *Exploring Life* (New York: D. Appleton, 1926), p. 167.
      Reis is as quoted by Thompson, p. 5.
140 "To all such clap-trap as this . . .": ibid., p. 37.
      "Besides the human voice . . .": ibid., p. 86.
      "found them perfectly competent to transmit speech . . .": ibid., p. 47.
141 Peter's description of the test is from Thompson, p. 127.
142 Mackay on Boursel is Mackey, p. 89.
      The Noad receiver is mentioned by Aitken, p. 35.
143 Make-and-break connections: Silvanus Thompson would argue, with much justification, that the resistance of an imperfect point of contact does vary continuously. When small particles are jostled together, for example, the electrical resistance can change by orders of magnitude. How exactly this happens was not understood at the time but is the principle of early carbon microphones, for example, and the coherer (chapter 14). Reis did indeed publish diagrams of oscillatory waves caused by speech, though he did not explain how his telephone reproduced such oscillations. Moreover, in 1971 Reis' instrument was tested at the Smithsonian and the resistance did vary continuously for low volumes, as one might expected of a coherer. (See Elliot Sivowitch in Shiers.) All of which goes to show that some questions are better decided in a laboratory than a courtroom. See also note to p. 144 below.
144 For Watson's yarn, see Watson, p. 78. For more on it, see Bruce, pp. 181–82.
      Bell's patent application is reproduced in full by Coe, Appendix 10, and is also in Bell's *Deposition*, pp. 451–61.
      Bell is rather vague about when he learned of Reis. See the *Deposition*, pp. 428–29. His account of his visit with Henry is pp. 46–47.

Regarding Reis' anticipation of Bell, in an interesting passage widely quoted by opponents of Bell but not by Bell biographers, Bell is being deposed at the Patent Office:

> *Question 37.* "If a Reis Telephone, made in accordance with the descriptions published before the earliest dates of your invention, would in use transmit and receive articulate speech as perfectly as the instruments which were used by you on June 25, 1876, at the Centennial, would it be proof to you that such Reis Telephones operated by the use of undulatory movements of electricity in substantially the same way as your instruments did upon the occasion referred to?
>
> *Answer by Bell.* "The supposition contained in the question cannot be supposed. Were the question put that if I were to hear an instrument give forth articulate speech transmitted electrically as perfectly as my instruments did on the occasion referred to in the question, I would hold this as proof that the instrument had been operated by undulatory movements of electricity, I would unhesitatingly answer, Yes."

From "Evidence for A. G. Bell," *Proceedings of the United States Patent Office Before the Commissioner of Patents,* p. 14, as quoted by Thompson, p. 178. See also Coe, p. 20.

145 Gray's caveat is quoted extensively by Aitken, chap. 8.

146 "Gray had not put his ideas to any practical test": Mackay, p. 124.
"What is remarkable is that Gray . . .": Mackay, p. 125.
"During his meeting with Wilber . . .": Mackay, p. 126.
"During this period I was not in Washington . . .": Bell, *Deposition,* p. 434.

147 "At the time, Bell did not know how . . .": Coe, p. 5.
"Gray's chagrin was intensified . . .": Mackay, p. 197.
Wilber's two affidavits, and Bell's, are reprinted by Coe in Appendix 7.

148 "Not until late fall of 1875 did Gray . . .": Bruce, p. 169.
"I believe the discovery of the true method . . ." is Frank L. Pope, as quoted by Aitken, p. 62. It is not entirely clear which lecture of Gray he is referring to. Probably he means "On the Transmission of Musical Tones Telegraphically," *Journal of the American Electrical Society* **1,** pp. 3–15 (1875), but more germane is "On Some Phenomena Attending the Transmission of Vibratory Currents of Electricity," *Journal of the American Electrical Society* **2,** pp. 69–81 (1878).
For Edison's priority in the liquid variable-resistance device, see Aitken, pp. 64, 86.
All the patents mentioned are available from the U.S. Patent Office.

149 Gray's letter can be found in the *Telephone Suits,* p. 151, or in Bruce, pp. 221–23.
For Drawbaugh's testimony, see Bruce, pp. 274–75, or Mackay, pp. 193–94.

150 Bruce on Dolbear is p. 277. Coe's account is pp. 41–43. For the courts' decisions, see Aitken, p. 59.

151 For Edison's contribution to the telephone, see Neil Baldwin, *Edison: Inventing the Century* (New York: Hyperion, 1993), pp. 72–73.
For more on Berliner, see Coe, pp. 30–35, and Bruce, p. 262.
Bell's contributions to the phonograph are documented by Leslie Newville, "Development of the Phonograph at Alexander Graham Bell's Volta Laboratory," in *Contributions from the Museum of History and Technology,* Bulletin 218, United States National Museum (Washington, D.C.: Smithsonian Institution Press, 1959). See also Bruce, pp. 250–54, 350–54.

152 Gray's epitaph is as cited by Lloyd, p. 251, who claimed to have the original in his possession.

14. A Babble of Incoherence: The Wireless Telegraph, a.k.a. Radio

An excellent though moderately technical history of early radio is Hugh Aitken, *Syntony and Spark: The Origins of Radio* (Princeton: Princeton University Press, 1985).

The definitive work about Lodge and his contributions is now Peter Rowlands and J. Patrick Wilson, eds., *Oliver Lodge and the Invention of Radio* (Liverpool: PD Publications, 1994). See in particular chaps. 6 and 10.

For a "neutral" view of the dispute between Lodge and Marconi, see Sungook Hong, "Marconi and the Maxwellians: The Origin of Wireless Telegraphy Revisited," *Technology and Culture* **35**, pp. 717–49 (1994).

A very useful account of coherers, etc., is Vivian J. Phillips, *Early Radio Detectors* (London: Peter Peregrinus, 1980). See especially the introduction and chaps. 1–3.

Still helpful after a century is John Fahie, *A History of Wireless Telegraphy* (Edinburgh: William Blackwood and Sons, 1900).

For Marconi, see:

Orrin E. Dunlap Jr., *Marconi, the Man and His Wireless* (New York: Macmillan, 1937).
W. P. Jolly, *Marconi* (London: Constable, 1972).
W. J. Baker, *A History of the Marconi Company* (London: Metheun, 1970).
This history of the Italian navy coherer is largely based on Vivian J. Phillips, "The Italian Navy Coherer Affair," *IEE Proc-A* **140**, pp. 175–85 (1993).
An evenhanded biography of Bose is Subrata Dasgupta, *Jagadish Chandra Bose* (New Delhi: Oxford University Press, 1999).

See also:

Probir K. Bondyopadhyay, "Sir J. C. Bose's Diode Detector Received Marconi's First Transatlantic Wireless Signal," *Proc. of the IEEE* **86**, pp. 259–85 (1998).
Probir K. Bondyopadhyay, "Under the Glare of a Thousand Suns—The Pioneering Works of Sir J. C. Bose," *Proc. of the IEEE* **86**, pp. 218–24 (1998).
D. T. Emerson, "The Work of Jagadis Chandra Bose," http://www.tuc.nrao.edu/~demerson/bose/bose.html.

For Popov, see:

Charles Susskind, "Popov and the Beginnings of Radiotelegraphy," *Proc. of IRE* **50**, pp. 2036–47 (1962).
James Rybak and Leonid Kryzhanovsky, "Alexander Popov: Father of Russian Wireless Telegraphy," *Old Timer's Bulletin* **41**, pp. 25–29 (2000).

153 For the letters, see *The New Scientist,* letters, June 10 and July 1, 1995. See also Susan Aldridge, "A Prizefight on the Wireless," May 20, 1995, pp. 46–47.

154 For an amusing account of the improvement of memory, see Aitken, pp. 115–23 and footnote 70, p. 174. For a detailed investigation see Hong.
Lodge's own account is in his autobiography *Past Years* (New York: Scribner's, 1932), chap. 17. See Rowlands and Wilson for a detailed discussion.
For Onesti, see Camillo Olivetti, "The Invention of the Coherer," *Electrical World and Engineer* **34**, pp. 858–59 (1899). This is an abridged translation and explanation of Onesti's original papers, which appeared in *Il Nuvo Cimento,* October 15, 1884, and March 2, 1886.

155 Righi's article is in *Il Nuovo Cimento* **35**, pp. 12–17 (1894).
Righi to Lodge is as quoted in Rowlands and Wilson, pp. 94, 158.

156 Lodge to the *Times* is as quoted by Jolly, p. 46.

157 The *Electrician* editorial is as quoted by Aitken, p. 208.

For Preece's role in the affair, see David Sealey, "Marconi Waves," in Rowlands and Wilson. Marconi's statement appeared in *McClure's Magazine* (see note to pp. 161–62).

157 For the "strong evidence," see Rowlands and Wilson, p. 95; Hong, p. 741.

158 The most complete discussion of the tuning patents is in Aitken, pp. 247–53.

160 All the quotations in the Italian Navy Coherer scandal are as in Philips.

161 J. C. Bose, "On a Self-recovering Coherer and the Study of the Cohering Action of Different Metals," *Proc. Roy. Soc. Lon.* **65,** pp. 166–72 (1899).

162 The back-to-back interviews with Bose and Marconi are in *McClure's Magazine,* March 1897, pp. 383–92. They are reprinted in Bondyopadhyay, "Sir J. C. Bose's Diode Detector."
Marconi's patent can be found in Fahie, p. 308.

164 Popov's paper is "Pribor' dlya obnaruzhenia i registrirovania elektricheskikh kolebania," *Journal of the Russian Physico-Chemical Society* **28,** 1, pp. 1–14 (1896). Popov is as quoted by Susskind, p. 2040, and Rybak and Kryzhanovsky, p. 28.
A detailed account of the Soviet claims is in Susskind.

165 For some words on Tesla, see Marc Seifer, "Reconstructing Tesla," *Scientific American,* April 1997, at the *Scientific American* Web site, www.sciam.com.
For typical received wisdom on Tesla, see "Tesla: Life and Legacy" at the PBS Web site (www.pbs.org/tesla).
William Broad, "Tesla, a Bizarre Genius Regains Aura of Greatness," *New York Times,* August 28, 1984, p. C1.
A typical Teslite biography is Margaret Cheney, *Telsa: A Man Out of Time* (New York: Dorset Press, 1989).

166 For the Supreme Court decision, see Aitken, p. 258, and "Misreading the Supreme Court: A Puzzling Chapter in the History of Radio," *Antenna,* Nov. 1998, http://www.mercurians.org/nov98/misreading.html.
For Rutherford, see Rowlands and Wilson, pp. 79–80, 149–50.
"There may never be another genius . . ." is Dunlap, p. 3.
Sir Bernard Lovell's evaluation is from his review of *Oliver Lodge and the Invention of Radio* in *Notes Rec. Roy. Soc. Lon.* **51,** pp. 151–53 (1997).

167 For Ambrose's recollections and Marconi's daughter, see note to p. 154.

15. Mind-Destroying Rays: Television

The two standard histories of the television are by Abramson. These works are crammed with facts and technical details, but the author provides virtually no connective tissue and the average reader will find them, in particular the first, quite daunting. They are:

> Albert Abramson, *The History of Television 1880 to 1941* (Jefferson, N.C.: McFarland, 1987).
>
> Albert Abramson, *Zworykin: Pioneer of Television* (Urbana: University of Illinois Press, 1995).
>
> A collection of original technical articles and histories by pioneers in the field is George Shiers, ed., *Technical Development of Television* (New York: Arno Press, 1977).

For more on Rozing, see:

> "Inventors of Television: Boris Rosing," *Radio Electronics,* Apr. 1966, p. 62;
>
> James Rybak, "Boris Rozing, Electronic Television Visionary," *Old Timer's Bulletin* **41,** 3, pp. 24–27, 34 (2000).

For more on Campbell-Swinton, see:

> T. H. Bridgewater, *A. A. Campbell Swinton, F.R.S.* (London: Royal Television Society, 1982).

Much can be found on the Internet about the Farnsworth-RCA patent battle. One detailed account from the Farnsworth point of view is at www.farnovision.com.

169 *Time*'s remark is in *Time* magazine, Dec. 31, 1999, "Person of the Century," p. 57.

171 Campbell-Swinton's letter can be found in full in Bridgewater, p. 20, or Abramson, *History*, pp. 28–29.

172 Rozing's statement is from P. K. Gorokhov, "History of Modern Television," in Shiers, p. 74.

174 More on Theremin's contributions to television can be found in Albert Glinsky, *Theremin: Ether Music and Espionage* (Urbana: University of Illinois Press, 2000).

178 For *Time*'s view of the Farnsworth-RCA feud, see *Time* (March 29, 1999), pp. 92–94.

16. Plausibility: The Invention of Secret Electronic Communication

There is no reference that covers the "intellectual" genesis of the Lamarr-Antheil patent. For Antheil's brief description of the adventure, see his autobiography, *Bad Boy of Music* (New York: Samuel French, 1990 [1945]), especially chap. 32.

Hedy Lamarr's exploits may be found in her autobiography, *Ecstasy and Me* (New York: Fawcett, 1966).

For more about the influence of the fourth dimension on twentieth-century art, see Linda Henderson's *The Fourth Dimension and Non-Euclidean Geometry in Modern Art* (Princeton: Princeton University Press, 1983), p. 328, which is a comprehensive treatment of this fascinating subject.

179–80 Antheil's description of his meeting with Hedy Lamarr can be found in *Bad Boy*, chap. 32.

181 More about Pound's opera can be found in Humphrey Carpenter's *A Serious Character: The Life of Ezra Pound* (New York: Delta, 1988).

181–82 The quotes about music and machines are from *Antheil and the Treatise on Harmony* (New York: Da Capo Press, 1968). See especially pp. 50–58.

182 The excerpt from Antheil's manifesto in *De Stijl* can be found in Henderson, p. 328.

Henderson also discusses the *Scientific American* contest. The first-prize essay was published in *Scientific American,* July 3, 1909, pp. 6, 15. All the winning essays are still available in Henry Parker Manning, ed., *The Fourth Dimension Simply Explained: A Collection of Essays Selected from Those Submitted in the* Scientific American*'s Prize Competition* (New York: Munn & Co., 1910).

184 Antheil's See Note system was showcased in "How to Play Two-Handed Piano," *Esquire,* January 1938, p. 52, and "6 Sharps That Beat in 3/4 Time," *Esquire,* March 1938, p. 106.

Antheil's handbook on hormonal criminology is *Every Man His Own Detective: A Study of Glandular Criminology* (New York: Stackpole, 1937).

Antheil's novel under the pseudonym Stacey Bishop is *Death in the Dark* (London: Faber and Faber, [c. 1930]). It is now unavailable except on microfilm from the University of California, Los Angeles.

185 Antheil's *Esquire* articles on endocrinology are "Glands on a Hobby Horse" (April 1936, p. 47); "Handbook for the Questing Male" (May 1936, p. 40); "The Glandbook in Practical Use" (June 1936, p. 36).

186 The information on Mandl, as well Antheil's dealings with the navy, is from Hans-Joachim Braun, "Advanced Weaponry of the Stars," *American Heritage of Invention and Technology,* spring 1997, p. 10.

189 The *Stars and Stripes* interview was in the Nov. 19, 1945, edition, reprinted in Antheil's *Bad Boy*.

The author is grateful for Hans-Joachim Braun's information on the document he found in the Bundesarchiv/Militararchive. It is apparently no longer available to him. The letters from Antheil to Reynolds and Lamarr are in the Antheil Collection at Columbia University's Rare Book and Manuscript Library. My thanks to Mauro Piccinini for making them available to me.

190 The history of SIGSALY and spread-spectrum technology is based on David Kahn's "Cryptology and the Origins of Spread Spectrum," *IEEE Sprectrum,* September 1984, p. 70, and on the articles on the NSA Web site, www.nsa.gov.
I wish to thank Robert Price for his extensive information on the history of secret communication.

III. The Domain of Chemistry and Biology

Introduction

Two good surveys of early chemistry are Henry M. Leicester, *The Historical Background of Chemistry* (New York: Dover, 1956), and Aaron J. Ihde, *The Development of Modern Chemistry* (New York: Dover, 1964).

For biographies of the philosophers mentioned:
Boris Menshutkin, *Russia's Lomonosov* (Princeton: Princeton University Press, 1952).
Mitchell Wilson, "Court Rumford," *Scientific American,* October 1960, pp. 158–68; Sanborn Brown, *Benjamin Thompson, Count Rumford* (Cambridge, Mass.: MIT Press, 1979). Wilson's article is a good summary; however, he repeats several stories that according to Brown have no basis in fact; see note to pp. 195–97 below. James Hofmann, *Andreé-Marie Ampère* (Oxford: Blackwell, 1995).

194 Antoine Lavoisier, *Elements of Chemistry,* trans. Robert Kerr (New York: Dover, 1965), p. 96. Lomonosov is as quoted by Menshutkin, p. 117. The allegations that Lavoisier knew of Lomonosov's work can be found in Leicester, p. 143.

197 Rumford's remarks on cannon boring are from Brown, p. 197.
Numerous legends surround the improbable career of Benjamin Thompson. One is that Thompson walked thirty miles a day to attend lectures at Harvard, where he learned physics from the famous John Winthrop. (He may have attended one or two, but Winthrop stopped lecturing for the summer a few days after Thompson started attending.) Another is that Thompson left England to fight in the colonies because of accusations that he was spying for the French (there were rumors), and a third is that he established the mechanical equivalent of heat (the number of calories in one joule of energy) decades before anyone else. This last story is supposedly based on a passage from Joule himself, but according to Brown, Joule never wrote anything of the sort.
For the remarks on the Rumford-Lavoisier union, see Brown, pp. 285, 289.

198 For Ampère's gravestone, see Hofmann, p. 367.

199 Avogadro's paper in translation is available on the Web at http://webserver.lemoyne.edu/faculty/giunta/avogadro.html.

200 For more on the periodic table, see Ihde; also Eric Scerri, "The Evolution of the Periodic System," *Scientific American,* September 1998, pp. 78–83.

201 Vladimir Vernadsky, *The Biosphere,* trans. David Langmuir (New York: Copernicus/Springer-Verlag, 1998).

17. The Evolution of Evolution: Erasmus, Charles, Gregor, and Ronald

For this chapter I am heavily indebted to:
>Desmond King-Hele, *Erasmus Darwin* (London: Giles de la Marc, 1999).
>Desmond King-Hele, ed., *The Essential Writings of Erasmus Darwin* (London: MacGibbon & Kee, 1968).

Also very helpful have been:
>George B. Dyson, *Darwin Among the Machines* (Reading, Mass.: Perseus, 1997), chap. 2.
>Janet Browne, *Charles Darwin, Voyaging* (Princeton: Princeton University Press, 1995).
>Cyril Darlington, *Darwin's Place in History* (Oxford: Basil Blackwell, 1959), p. 33.

203 Regarding creationism in the school system: For several years I taught calculus from a widely used text by Varverg and Purcell. In the eighth edition of a photograph of the Shroud of Turin is accompanied by a caption stating that radiocarbon dating has determined that the shroud was created in the fourteenth century and therefore could not be the burial shroud of Christ. In the ninth edition the photograph and caption have been removed.

204 Seward's verse is as published by King-Hele, *Erasmus,* p. 89.

205 Genista is in *Loves* (I, pp. 57–64), as found in King-Hele, *Essential,* p. 143.

206 *Temple of Nature* (I, pp. 296–303, and pp. I, pp. 247–50) as quoted in *Essential,* p. 90.

207 *Zoonomia* II, p. 240, as quoted in *Essential,* p. 87.

208 Charles Darwin, *Autobiography,* ed. Nora Barlow (London: Collins, 1958), p. 49.
Darwin to Hooker is from 1844, as quoted by Barlow in *Autobiography,* p. 160.
Darwin's footnote from the "Historical Sketch" is as quoted by Darlington, p. 33.

209 For Butler on Darwin and Darwin's reply, see Samuel Butler, *Unconscious Memory* (London: A. C. Fitfield, 1920), chap. 4. See also Dyson, chap. 2, and Barlow's appendix to the *Autobiography.*

210 Darwin to Huxley is from June 1859, as quoted by Darlington, p. 33.
For Blyth, see Loren Eiseley, "Charles Darwin and Edward Blyth," *Proceedings of the American Philosophical Society* **103,** pp. 94–158 (1959). Darwin to Lyell on Lamarck is October 1859, as quoted by Eiseley, p. 111.
"As in brute creation . . .": Edward Blyth, "An attempt to classify the 'varieties' of animals, with observations on the marked seasonal and other changes which naturally take place in various British species, and which do not constitute varieties," *Magazine of Natural History* **3,** pp. 40–53 (1835). Reprinted as Appendix A to Eiseley; see p. 119.

211 For more on Lawrence and other neglected precursors, see Darlington. The summation of Lawrence's views is largely as given there on pp. 19–20. Lawrence's book *Lectures on Physiology, Zoology and the Natural History of Man* is extremely rare. The quotations are from the 1822 edition housed in the Bryn Mawr College Rare Book Room, pp. 249 and 225, respectively.

212 The story of Mendel's "neglect" is based on Conway Zirkle, "Some Oddities in the Delayed Discovery of Mendelism," *Journal of Heredity* **55,** pp. 65–72 (1964). See also Eugene Garfield, "Would Mendel's Work Have Been Ignored if the Science Citation Index Was Available 100 Years Ago?" *Essays of an Information Scientist* **1,** pp. 69–70 (1962).

213 A splendid annotated version of Mendel's paper is available on the Web at http://www.netspace.org/MendelWeb/Mendel.html.

214 Thomas Knight's paper is *Phil. Trans. Roy. Sec. Lon.* **89,** pp. 195–204 (1799). John Goss is *Trans. Horticultural Soc. Lon.* **5,** pp. 65–72 (1824). Dzierzon and the other pre-Mendelians are as discussed in Zirkle, *Isis* **42,** pp. 97–104 (1951).

For more on Darwin's beliefs on the origin of variability, see Theodosius Dobzhansky, "Variations and Evolution," *Proceedings of the American Philosophical Society* **103,** pp. 252–63 (1959).

Ronald Fisher, "Has Mendel's Work Been Rediscovered?" *Annals of Science,* **1,** pp. 115–37 (1936), p. 132. This paper is available online at www.library.ade/aide.edu.au/digitised/fisher/genetics.html.

18. Dreams with Open Eyes: Kekulé, Benzene, and Loschmidt

The writings of Alfred Bader and his colleagues have been the primary source material for this chapter. See:

Alfred Bader, "Joseph Loschmidt—The Father of Molecular Modelling," in *Adventures of a Chemist Collector* (London: Weidenfeld and Nicolson, 1995).

William J. Wiswesser, "Johann Josef Loschmidt: A Forgotten Genius," *Aldrichimica Acta* **22,** p. 17 (1989).

Christian Noe and Alfred Bader, "Facts Are Better than Dreams," *Chemistry in Britain,* February 1993, p. 126.

Alfred Bader and Leonard Parker, "Joseph Loschmidt, Physicist and Chemist," *Physics Today,* March 2001, www.physicstoday.org/pt/vol-54/iss-3/p45.html.

216 The excerpt of Kekulé's speech is as given by Royston Roberts in *Serendipity* (New York: John Wiley and Sons, 1989). This translation is the one by Francis R. Japp, originally published in *J. Chem. Soc. Lon.* **73,** p. 97 (1898). It can also be found, with several alternative translations, in John H. Wotiz, ed., *The Kekulé Riddle* (Carbondale, Ill.: Glenview Press, 1993), chap. 19.

219 Much of the information about Couper is found in Leonard Dobbin's "The Couper Quest," *J. Chem. Educ.* **11,** p. 331 (1934). This article contains interesting correspondence of Anschütz and Crum Brown relating to their hunt for Couper, including the comment that "Couper was a complete wreck." See also Richard Anschütz, "The Life and Work of Archibald Scott Couper," *Proc. Roy. Soc. Edin.* **29,** p. 193 (1909). One finds contradictory statements about whether Couper introduced the quadrivalence of carbon independently of Kekulé. Anschütz, on the one hand, says, "In Couper's paper 'On a new chemical theory,' the hypothesis of the quadrivalence of carbon and the concatenation of carbon atoms was developed shortly after the same had been done by Kekulé, and it was—as I have already indicated—only by an accident fateful for Couper that his paper did not appear simultaneously with that of Kekulé" (p. 200). On the other hand, one page later Anschütz says, "Couper does not claim for himself the hypothesis of the quadrivalence of carbon" (p. 201). In Couper's original paper, *Annales de chimie et de physique* **53,** p. 469 (1858), the hypothesis of quadrivalence is clearly stated, with no reference to anyone else.

For the claim of Odling's priority in the tetravalence of carbon, see C. A. Russell, *The History of Valency* (New York: Humanities Press, 1971), p. 120.

For Kekulé's attack on Couper, see Anschütz; Russell, p. 125; or "Sausage and Structural Formulae," http://www.chem.yale.edu/~chem125/125/history/Kekule/Kekule.html.

220 Kekulé's relationship to Loschmidt and his letter to Ehrlenmeyer are discussed in Richard Anschütz, *Auguste Kekulé* (Berlin: Verlag Chemie, 1929), pp. 296–305.

220–21 Letters of Borodin to his wife and evidence for fabrication of the dream story can be found in Wotiz, chap. 17. This includes the excerpt from Kopp's book.

222 For an anti-Bader view, see Alan J. Rocke, "Waking up to the Facts?" *Chemistry in Britain,* May 1993, p. 401.

223 For much more on the direction of time, see Tony Rothman and George Sudarshan, *Doubt and Certainty* (Reading, Mass.: Perseus Books, 1998), chap. 5.

19. Chance, Good and Bad: Penicillin

A detailed account of the penicillin story is Gwyn Macfarlane, *Alexander Fleming: The Man and the Myth* (Cambridge, Mass.: Harvard University Press, 1984).

The authorized biography is André Maurois, *Life of Alexander Fleming* (New York: E. P. Dutton, 1959).

225 "There is something deeply moving . . .": Maurois, p. 154.
Maurois' account of the discovery is on p. 127.

226 Ian Curteis' dialogue is from "Life of Sir Alexander Fleming," BBC, 1972, as quoted by Macfarlane, p. 142.
George Papacostas and Jean Gaté, *Les Associations microbiennes* (Paris: Gaston Doin, 1928). See also discussion by Macfarlane, pp. 135–36.
J. Burdon-Sanderson's researches can be found in "Dr. Sanderson's Further Report of Researches concerning the Intimate Pathology of Contagion," *The 13th Report of the Medical Officer of the Privy Council, 1870* (London: HMSO, 1871).

227 William Roberts, *Proc. Phil. Trans. Roy. Soc. Lon.,* **164,** p. 457 (1874).

228 Tyndall's remarks on the greenhouse effect are from "On the Absorption and Radiation of Heat by Gases and Vapours, and on the Physical Connexion of Radiation, Absorption, and Conduction," *Proc. Royal Soc.* **9,** pp. 100–4 (1861), as quoted in W. H. Brock, N. D. McMillan, and R. C. Mollan, eds., *John Tyndall: Essays on a Natural Philosopher* (Dublin: Royal Dublin Society, 1981), p. 91.

228–29 For Tyndall on penicillin, see John Tyndall, "The Optical Deportment of the Atmosphere in Relation to the Phenomena of Putrefaction and Infection," *Phil. Trans. Roy. Soc. Lon.* **166,** p. 27 (1876).
Pasteur's remarks can be found in *Ouvres de Pasteur Reunies* par Pasteur Vallery-Radot (Masson et cie. 1922–1939), vol. 6, p. 177. The original is in the Comptes Rendus for the session of July 16, 1877, **85,** pp. 101–5.

230 Duchesne's thesis is available in French on the Internet at http://perso.wanadoo.fr/jdtr/struc/duchesne.htm.
"In this case, too, the search was not continued": Maurois, p. 129.

IV. The Domain of Mathematics: Closed for Renovation

For Galois, see Tony Rothman, "Genius and Biographers: The Fictionalization of Evariste Galois," in *Science à la Mode* (Princeton: Princeton University Press, 1989) or on the author's Web site. See also Laura Toti Rigatelli, *Evariste Galois* (Boston: Birkhaüser, 1996).

For Cartwright and Littlewood, see Shawnee McMurran and James Tattersall, "The Mathematical Collaboration of M. L. Cartwright and J. E. Littlewood," *American Mathematical Monthly* **103** (1996), pp. 833–45. Cartwright and Littlewood's original paper on what is now called period doubling is *J. of Lon. Math. Soc.* **20,** pp. 180–95 (1945). (The important reference in the *AMM* article is incorrect.)

For the (lack of the) Nobel prize in math, see Elisabeth Crawford, *The Beginnings of the Nobel Institution* (Cambridge, Mass.: Cambridge University Press, 1984), p. 113.

INDEX